# FORMS OF MATHEMATICAL KNOWLEDGE

*Learning and Teaching with Understanding*

*Edited by*

DINA TIROSH

*School of Education, Tel Aviv University, Israel*

Reprinted from *Educational Studies in Mathematics* 38 (1–3), 1999

KLUWER ACADEMIC PUBLISHERS

DORDRECHT / BOSTON / LONDON

Library of Congress Cataloging-in-Publication Data

A. C.I.P. Catalogue record for this book is available from the Library of Congress

ISBN 0-7923-5995-X

---

Published by Kluwer Academic Publishers,
P.O. Box 17, 3300 AA Dordrecht, The Netherlands.

Kluwer Academic Publishers incorporates
the publishing programmes of
D. Reidel, Martinus Nijhoff, Dr W. Junk and MTP Press.

Sold and distributed in the U.S.A. and Canada
by Kluwer Academic Publishers,
101 Philip Drive, Norwell, MA 02061, U.S.A.

In all other countries, sold and distributed
by Kluwer Academic Publishers Group,
P.O. Box 322, 3300 AH Dordrecht, The Netherlands.

*Printed on acid-free paper*



Printed in the Netherlands

# TABLE OF CONTENTS

DINA TIROSH / Introduction 1

EFRAIM FISCHBEIN / Intuitions and Schemata in Mathematical Reasoning 11

DINA TIROSH and RUTH STAVY / Intuitive Rules: A Way to Explain and Predict Students' Reasoning 51

PAUL ERNEST / Forms of Knowledge in Mathematics and Mathematics Education: Philosophical and Rhetorical Perspectives 67

TOMMY DREYFUS / Why Johnny Can't Prove 85

EDDIE GRAY, MARCIA PINTO, DEMETRA PITTA and DAVID TALL / Knowledge Construction and Diverging Thinking in Elementary & Advanced Mathematics 111

JOHN MASON and MARY SPENCE / Beyond Mere Knowledge of Mathematics: The Importance of Knowing-To Act in the Moment 135

THOMAS J. COONEY / Conceptualizing Teachers' Ways of Knowing 163

ANNA O. GRAEBER / Forms of Knowing Mathematics: What Preservice Teachers Should Learn 189

PESSIA TSAMIR / The Transition from Comparison of Finite to the Comparison of Infinite Sets: Teaching Prospective Teachers 209

RUHAMA EVEN / Integrating Academic and Practical Knowledge in a Teacher Leaders' Development Program 235

# FORMS OF MATHEMATICAL KNOWLEDGE: LEARNING AND TEACHING WITH UNDERSTANDING

*This volume is dedicated*
*to the memory of Professor Efraim Fischbein,*
*a distinguished member*
*of the mathematics education community*

This volume has its origin in the presentations and discussions that occurred in the working group on 'Forms of Mathematical Knowledge' of the Eight International Conference on Mathematical Education (ICME-8) held in Seville, Spain, in July 1996. The working group was mainly devoted to defining, discussing and contrasting psychological and philosophical issues related to various types of knowledge involved in mathematics learning and teaching (e.g., knowing that, knowing how, knowing why and knowing to; intuitive and analytical; explicit and tacit; elementary and advanced). After the conference the participants felt that issues raised in the working group were significant for the mathematics education community and decided to submit a proposal for a special issue on forms of mathematical knowledge to *Educational Studies in Mathematics.*

This introduction attempts to provide the reader with a preliminary flavor of the volume. It describes its structure and presents a brief synopsis of each article.

## STRUCTURE

This volume consists of two parts:

### *PART I: interactions between various forms of mathematical knowledge*

Various forms of knowledge have been frequently dealt with in the mathematics education literature e.g., instrumental, relational, conceptual, procedural, algorithmic, formal, visual, intuitive, implicit, explicit, elementary, advanced, knowing that, knowing why and knowing how. The articles

*Educational Studies in Mathematics* **38:** 1–9, 1999.

in this part define various forms of mathematical knowledge, describe the distinctions between them and discuss possible implications of various forms of mathematical knowledge for mathematics learning and teaching.

*Part II: Forms of Teachers' mathematical knowledge*

Teaching is increasingly being recognized as a difficult profession that requires decision making and problem solving in a public, complex and dynamic environment for many hours daily. There is growing recognition that mathematical knowledge alone does not guarantee better teaching and attempts are being made to define the various forms of knowledge needed for teaching. The articles in this part are devoted to differentiating and characterizing forms of knowledge needed for teaching, ideas about forms of mathematical knowledge that are important for teachers to know, and ways of implementing such ideas in preservice and professional development teaching programs.

## Summaries of Articles

*PART I: interactions between various forms of mathematical knowledge*

This part includes six articles. Articles 1–3 deal with issues related to intuitive knowledge. These articles emphasize that although explicit propositional knowledge is important in both mathematics and mathematics education, other types of knowledge are essential as well.

In the first article, on *Intuitions and schemata in mathematical reasoning*, the late *Efraim Fischbein* analyzes the complex relations between intuitions and structural schemata. In previous work Fischbein defined the concept of intuitions, classified various types of intuitions and described the contribution, sometimes positive and sometimes negative, of intuitions in the history of science and mathematics and in the didactic process. In this article Fischbein goes one step further to explain the shaping and functioning mechanisms of intuitions, which operates subconsciously. In the introduction, Fischbein argues that knowledge about intuitive interpretations is crucial to teachers, authors of textbooks and didactical researchers alike, because formal knowledge very often collides with intuitive interpretations, which are naturally very resistant to and conflicting with scientifically established notions. The first part of the article is devoted to the concept of intuition, and Fischbein describes the general characteristics of intuitive cognitions, classifies intuitions, discusses the correctness of intuitions, distinguishes amongst various types of relationships between intuitions, and draws some didactic implications. The second part of the article deals with the concept of schema. Fischbein presents various definitions and interpretations of this concept, distinguishes between structural schemata (general, basic ways of thinking and solving, related to the intellectual development of a person) and specific schemata (e.g., the algorithms of addition, the formula for solving a quadratic equation, etc.). In the third part Fischbein argues that intuitions are generally based on structural schemata: when an intellectual schema changes, as an effect of age and experience, corresponding intuitions change too. The transition from a schema, which is a sequential process, to an intuition which is a global, apparently sudden cognition, is achieved by a compression process. The article is full of fascinating examples illustrating the relations between intuitive cognitions and formal knowledge and is, as one of the reviewers noted, 'the culmination of a life's work, collating research and theory in intuition'.

The second article, on *Intuitive rules: A way to explain and predict students' reasoning*, by *Dina Tirosh* and *Ruth Stavy*, relates to possible common origins of students' intuitive reactions to mathematical and scientific tasks. The article suggests that students' responses to a given task are often determined by the specific formulation of this task and by a repertoire of intuitive rules. The main part of the paper is devoted to a description of one intuitive rule: 'Same A- same B' and demonstrates its explanatory and predictive power.

*Paul Ernest*, in the third article on *Forms of knowledge in mathematics and mathematics education: Philosophical and rhetorical perspectives,* concentrates on a dimension of intuitive knowledge that has not received enough attention in mathematics education, that is: tacit, rhetorical knowledge. The article opens with a short description of three relatively new developments in the history and philosophy of mathematics which have given rise to parallel developments in mathematics education: the emergence of the fallibilist philosophies of mathematics, the growing attention to the vital role of the social and cultural contexts in the creation and justification of mathematical knowledge, and the growing awareness of the significance of genres and rhetoric in mathematics. Ernest argues that these developments enhance the realization that tacit knowledge is important in mathematics, as in any other area of human thought, and describes different aspects of tacit knowledge of mathematics. A main part of Ernest's paper is devoted to ways of validating tacit knowledge, to the role of various types of tacit knowledge in the development and the assessment of both mathematics and school mathematics knowledge and to the need for explicit instruction related to the rhetorical styles of mathematics and school mathematics.

The fourth article is a natural sequel to the previous one. In *Why Johnny can't prove*, *Tommy Dreyfus* questions the extent to which it is reasonable to demand formal, written proofs from students at high school or even at a college level. The article opens with a sample of written explanations to given problems, illustrating that the task of explaining a given problem is extremely difficult even for those with reasonable proficiency and some understanding of this problem. A review of the research on high school and undergraduate students' conceptions of proof confirms that very few students ever learn to appreciate the characteristics of formal proofs and to construct such arguments themselves. A review of the research on epistemological and cognitive issues related to the concept of proof points out that (1) for mathematics educators there appears to be a continuum reaching from explanation via argument and justification to proof, but the distinctions between these categories are not sharp; and (2) the role, status and nature of proof is challenged in mathematics itself. Dreyfus argues that in light of this situation, it is hard to expect students to be able to distinguish between different forms of reasoning and to judge the validity of mathematical arguments. He lists several crucial issues regarding the criteria teachers and educators use to evaluate students' written explanations (e.g., On what basis do we or do we not accept a student's explanation? Which arguments do we accept under which circumstances; which don't we, and why? To what degree does an explanation need to convince? And if so, does it need to convince a mathematician, a teacher, fellow students?). Subsequently the author suggests that mathematics educators direct more efforts towards developing criteria teachers can use to judge the acceptability of their students' mathematical arguments. Dreyfus also calls for more research aimed at assessing changes in students' views of mathematics and in their ability to explain and justify.

In the fifth article, *Knowledge construction and diverging thinking in elementary and advance mathematics*, *Eddie Gray, Marcia Pinto, Demetra Pitta* and *David Tall* focus on the growth of mathematical knowledge from elementary to advanced mathematics. The authors introduce a theory of cognitive development in mathematics which is based on three essential components: object, actions and properties, and they discuss the similarities and differences between their theory and other theories and approaches concerning this issue. Gray et al. describe the differences between elementary and advanced mathematics and argue that the divergence between those who succeed in both elementary and advance mathematics and those who fail is rooted in the individual's ability to flexibly move between seeing a symbol as a process and as a concept . In respect to elementary arithmetic, Gray, Pinto, Pitta and Tall present evidence that children who succeed in elementary arithmetic are able to freely move between the objects counted, the process of counting, and the procept of numbers. Less successful children focus on the specific nature of the objects counted and associate them with real and imagined experiences. At a more advanced level, a student should understand that in mathematics the existence of objects with certain properties can be assumed, and on this basis the mental objects are reconstructed through formal proofs. At this level, less successful students are unable to focus on the generative power of definitions to construct the properties of the conceptual objects that are the essence of the formal theories. In the concluding part the authors offer several ideas to those who ask the obvious question: 'How can we help those using less successful methods of processing to become more successful?' and state that more research is needed to test the implications of the suggested theory of cognitive development for the teaching of elementary and advanced mathematics.

The last article in this part, by *John Mason* and *Mary Spence*, is entitled: *Beyond mere knowledge of mathematics: The importance of knowing-to act in the moment*. This article focuses on one form of knowledge – knowing to act in a moment. Mason and Spence first describe and discuss some traditional epistemological distinctions: knowing that, knowing how and knowing why. They argue that education driven by these three types of knowledge sees knowledge as a static object which can be passed on from generation to generation as a collection of facts, techniques, skills and approaches. These three forms of knowledge, which constitute knowing-about the subject, form the core of institutionalized education. Knowing-about, however, does not automatically develop the awareness that enables students to know to use this knowledge in new situations. A fourth form of knowledge, *knowing-to act* in the moment, is the type of knowledge that enables people to act creatively rather than merely react to stimuli with trained or habituated behavior. Mason and Spencer claim that knowing-to requires sensitivity to situational features and some degree of awareness of the moment, so that relevant knowledge is accessed when appropriate. They offer different approaches to what they mean by knowing-to act in the moment (experiential, experience-based and theoretical), define crucial aspects and characteristics of knowing-to (instantaneous, a gestalt, a sudden shift of the focus of attention) and describe mechanisms of shifts of attention (habituation, enculturation, generality, particularity, metonymic triggers and metaphoric structures). Mason and Spence's article is enriched with specific, vivid examples of instances of *knowing to* and offers suggestions of how to go about teaching in such a way as to help students know-to.

*PART II: Forms of mathematical knowledge: The teaching perspective*

Four articles are included in this part. While articles 7, 8 and 9 relate to prospective and inservice teachers, the last article (article 10) concerns teachers leader and inservice teachers educators. Article 7, by *Thomas Cooney*, on *Conceptualizing teachers' ways of knowing,* addresses issues relate to teachers' knowing mathematics and knowing about the teaching and learning of mathematics. The article opens with a review of the literature on what teachers know and believe about mathematics. Data gathered and analyzed from the Research and Development Initiatives Applied to Teacher Education (RADIATE) project are used to show that prospective teachers lack fundamental understandings of school mathematics despite their success in studying advanced university level mathematics. Moreover, both prospective and experienced secondary teachers tend to believe that good teaching is good telling, view clarity, pace of instruction and availability for answering questions as the characteristics of good teaching, tend to confuse between being the social authority in the classroom and the legitimizer of truth, and are limited in translating their mathematical knowledge into tasks that require a deep and through understanding of mathematics. Cooney argues that the orientation toward telling with clarity and the overwhelming propensity to be a caring teacher puts at risk ideas that may appear to contradict these characteristics, such as causing students to experience stress in solving problems.

In the second section of his article, Cooney describes four characterizations of teachers that seems promising as a way of conceptualizing the structure of teachers' beliefs: isolationist, naive idealist, naive connectionist and reflective connectionist. He provides evidence that the structure of one's beliefs is an important factor in determining what gets taught and how it gets taught, and argues that influencing the structure of teachers' beliefs ought to be an explicit focus of teacher education programs. In the third section Cooney describes an attempt to transform prospective secondary teachers' notions of telling and caring into a one that encourages attention to context and reflection, thereby creating a more student-centered classroom. He describes various interesting and challenging mathematical situations embedded in pedagogical contexts, that were proved to be effective in helping teachers develop a more relativistic view of mathematics. The fourth section raises moral dilemmas concerning mathematics teacher education. Cooney argues that the goal of teaching is to educate, and we, as teacher educators, have an obligation to enable teachers to see knowledge acquisition as power so that they can enable their students to acquire that same kind of power. In the concluding section, Cooney claims that forms of teachers' knowledge constitute an often neglected but critical element in the process of teacher education, and that conceptualizing forms of teacher knowledge and how that knowledge is held allows to move teacher education beyond an activity and toward an arena for disciplined inquiry.

In the eighth article, on *Forms of knowing mathematics: A teacher educator perspective*, *Anna Graeber*, focuses on important ideas about forms of knowing mathematics that teachers should meet in mathematics methods courses for preservice teachers. In the article, Anna Graeber describes five ideas that often challenge the preservice teachers' intuitive ideas about teaching and learning. The proposed ideas are related to Shulman's (1986) framework of teacher knowledge. Three ideas (understanding students' understanding is important, students knowing in one way do not necessarily know in the other(s), and intuitive understanding is both an asset and a liability), relate to the first cluster of pedagogical content knowledge, i.e., knowing what makes a subject difficult and what preconceptions students are likely to bring. The last two important ideas (certain characteristics of instruction appear to promote retention, and providing alternative representations and recognizing and analyzing alternative methods are important) relate to the second portion of pedagogical content knowledge, namely, knowing how to make the subject comprehensible to learners. Graeber draws some implications that each idea holds for teaching mathematics, and suggests interesting activities that were proved to be feasible in helping preservice teachers appreciate each of these ideas. Graeber concludes her article with two crucial issues for preservice teacher education: Are these the right big ideas? What are effective ways of helping preservice teachers learn ideas about pedagogical knowledge and how do they learn to apply them?

While article 8 relates to a method courses for prospective teachers, article 9 mainly focuses on the a mathematical course for the same population. *Pessia Tsamir*, in an article entitled: *The transition from comparison of finite to the comparison of infinite sets: Teaching prospective teachers*, presents a study that provided prospective secondary teachers with opportunities to experience some of the cognitive difficulties involved in progressing from one realm of numbers to an extended one and to critically reflect on this experience. In the first section of the article, Tsamir briefly describes an intuitive-based Cantorian Set Theory course that relate to students' tendencies to overgeneralize from finite to infinite sets, attempting to encourage prospective teachers to view infinite sets as an extension of finite sets, while discussing the similarities and differences between them. She then elaborates on one activity in which the participants were encouraged to reflect on their experiences (e.g., realizing their tendency to attribute properties of finite sets to infinite sets, describing apparent inconsistencies in their own thinking, relating to the role of intuitive beliefs in their thinking). The second section of this article is devoted to a through analysis of the impact of the intuitive-based mathematics course on prospective teachers' intuitive and formal knowledge of a central aspect of the Cantorian Set Theory, i.e., powers of infinite sets. In the concluding section, Tsamir discusses possible reasons for the success of the course, draws some conclusions to the learning and teaching of the Cantorian Set theory and cautiously suggests possible implications to other mathematical courses.

The concluding article, on *Integrating academic and practical knowledge in a teacher leaders' development program*, is one of a few that centers on the education of teacher leaders. *Ruhama Even* describes an attempt to make research in mathematics education meaningful for teacher leaders and inservice teacher educators. Her article focuses on the academic aspect of the Manor Program, a project which aims to develop a professional group of teacher leaders and inservice teacher educators whose role is to promote teacher learning about mathematics teaching. Even first briefly describes the Manor Program and details the methodology of her study. She reports on the participants' interest in learning about students' conceptions and ways of thinking through reading, presenting and discussing relevant research articles and also on their initial reluctance to conduct a mini-study on students' and teachers' ways of thinking and difficulties, replicating one of the studies presented in the course . Even shares with us the attempts of her project staff to encourage participants to conduct a mini-study and to write a report describing the subjects' ways of thinking and difficulties, comparing their results with the original study. She analyses the impact of the course on participants' understanding of students' and teachers' conceptions and ways of thinking about mathematics and the effectiveness of the use of the mini-studies as a means to encourage participants to examine knowledge learned in the academy in light of their practical knowledge and vice versa.

Ruhama Even found that participation in the course expanded teacher leaders and inservice teacher educators' understanding that students construct their own knowledge and ideas about the mathematics they learn in ways which are not necessarily intended by instruction. Conducting the mini-study was instrumental in making general, theoretical ideas more specific, concrete and relevant to participants and writing the report encourages reflective and analytical thoughts that supported intellectual restructuring. Although Even's study was not aimed at assessing the impact of promoting teachers' understanding of students' ways of thinking on their teaching practices, there were some initial indications that acquaintance with theoretical knowledge contributed to actual changes in participants' teaching practices and in their students learning. More research is needed to explore how participants use their knowledge about students' ways of thinking in their own practices, as teachers and as teachers leaders.

The last sentence in Ruhama Even's article discusses the importance of assisting teachers in making situational, intuitive and tacit practical knowledge more formal and explicit. This claim regarding teachers is nicely connected with the very first article in this special issue, in which Efraim Fischbein discusses the relationship between intuitive and formal knowledge, arguing that it is important for teachers to understand these complex interactions. So, as it often happens, the end is a new beginning.

## EPILOGUE

In bringing this introduction to a close, I would like to acknowledge the contribution of all authors who agreed to join me on this journey: to participate actively in the Working Group on 'Forms of Mathematical Knowledge' at ICME-8, to write thought-provoking articles, and to comment on other articles included in this volume. Special thanks go to Kenneth Ruthven and to Tommy Dreyfus who patiently and wisely guided me in this endeavor and were there for me whenever their help was needed.

This volume is dedicated to Efraim Fischbein, an excellent researcher, a great teacher and a cherished friend. Several years ago, when consulting with Efraim on the structure and organization of the Working Group, he suggested discussing several subjects, some of which are addressed in this volume and some not. A central issue that was not addressed concerns the role of various types of knowledge (intuitive, algorithmic and formal) in understanding specific mathematical domains (e.g., arithmetic, algebra, geometry, calculus, probability, topology). This, and many other issues related to 'Forms of Mathematical Knowledge' is of crucial importance to the learning and the teaching of mathematics with understanding. Moreover, in each of the articles the reader will find questions that need 'more attention', 'further exploration', or simply, 'more research'. The authors of this special volume would like it to stimulate and promote more discussion on forms of mathematical knowledge in the mathematics education community.

DINA TIROSH

EFRAIM FISCHBEIN†

# INTUITIONS AND SCHEMATA IN MATHEMATICAL REASONING

ABSTRACT. The present paper is an attampt to analyze the relationship between intuitions and structural schemata. Intuitions are defined as cognitions which appear subjectively to be self-evident, immediate, certain, global, coercive. Structural schemata are behavioral-mental devices which make possible the assimilation and interpretation of information and the adequate reactions to various stimuli. Structural schemata are characterized by their general relevance for the adaptive behavior. The main thesis of the paper is that intuitions are generally based on structural schemata. The transition from schemata to intuitions is achieved by a particular process of compression described in the paper.

## INTRODUCTION

The intuitive kind of knowledge has been a concept in which mainly philosophers have been interested. In the works of Descartes (1967) and Spinoza (1967) intuition is presented as the genuine source of true knowledge. Kant (1980) describes intuition as the faculty through which objects are directly known in distinction to understanding which leads to indirect conceptual knowledge. Bergson (1954) distinguishes between intelligence and intuition. Intelligence is the way by which one knows the material world, the world of solids, of space properties of static phenomena. By intuition we reach directly the essence of the spiritual life, the grasp of time phenomena ('la durée'), of motion. The Zeno paradoxes could be explained, in Bergson's view, by the fact that one tries to understand motion using the tools of intelligence which are not adequate for that aim. Motion has to be grasped in its genuine continuity, in its fluency, *directly*, through intuition. In his arguments, Zeno assumes that motion (and respectively, time) can be cut, divided, decomposed into tiny static pieces with which intelligence is able to operate. But the recomposition of motion from sequences of static components is an impossibility.

Some philosophers, like Hans Hahn (1956) and Bunge (1962) have criticized intuition and its impact on scientific reasoning: intuition may be misleading and, therefore, should be avoided in scientific reasoning.

Psychologists have shown and still show little interest in the theory of intuitive knowledge. Except for the books of Westcott (1968), Fischbein (1975, 1987) and Bastick (1982), there are no systematic monographs de-

*Educational Studies in Mathematics* **38:** 11–50, 1999.

voted to the domain of intuition. Noddings and Shore (1984) published an excellent book devoted to the relevance of intuition for pedagogy. The surprising fact is that in the usual textbooks of cognitive psychology, intuitive cognition is not even mentioned as a main component of our cognitive activity. I have not found one book on cognitive psychology in which a chapter is devoted to this topic (in the same line, for instance, with attention, memory, problem solving, etc.).

The interest in intuitive knowledge (though sometimes not under the same name) appears mainly in the work of people dealing with the development in children and students of scientific and mathematical understanding (see, for instance, the work of Clement et al., 1989; DiSessa, 1988; Gelman and Gallistel, 1978; McCloskey et al., 1983; Resnick, 1987; Stavy and Tirosh, 1996; Tirosh, 1991). This is only a very small sample of works done with regard to the practical and developmental aspects of intuition. As I said, they are inspired mainly by educational needs.

There is no commonly accepted definition for intuitive knowledge. The term 'intuition' is used as one uses mathematical primitive terms like point, line, set, etc. The meaning of intuition is implicitly considered as being intuitively evident. The commonly, implicitly accepted property is that of self-evidence, that is opposed to a logical-analytical endeavor.

The renewed interest in intuition in the last century has two main sources: One, is the growing endeavor of scientists – mathematicians and empirical scientists – to increase continuously the degree of rigor, of conceptual purity in their respective domains. Physicists aspire to reach the most elementary components of matter and the laws governing the respective forces and motions. On the other hand they tend to understand and explain the world as a whole, absolutizing our knowledge by taking into account the genuine relativistic nature of the physical laws. From Galileo to Newton, from Newton to Einstein, from Einstein to Planck, the image of the world has become more accountable in terms of consistency, rigor and objectivity. The basic tendency in modern times has been to 'purify' our knowledge from subjective, direct interpretations and beliefs and render it in accord with 'objective' data rigorously acquired. This led to increasing contradictions between what appears to be obvious and what one obtains as a result of a rigorous analysis of 'scientifically' acquired data.

In mathematics, until the 19th century, everything seemed to be more or less consistent. But the same efforts towards rigor led to the discovery of the same type of cognitive conflicts. It seemed self-evident that every continuous real function is differentiable in one point at least. But Weierstrass showed that it is not so. He offered the example of a function which does not admit any tangents at any of its points, but so had Bolzano also long before him. Until the 19th century, the Euclidean geometry was *The Geometry* based on apparently the same grounds of self-evident axioms. Lobachevsky, Bolyai, Riemann showed that other geometries are also logically possible. These non-Euclidean geometries conflict with our natural, apparently self-evident, image of the real world and its space properties.

Until the 19th century, the concept of actual infinity has been excluded from mathematics because one could easily prove that it leads to contradictions. Cantor, changing the point of view with regard to the comparison of the cardinals of sets, proved that actual infinity should be accepted as a non-contradictory concept in mathematics. But this formally non-contradictory interpretation of infinity conflicts with our natural, relatively spontaneous interpretations. And so on ...

The systematic striving for consistency and rigor in science and mathematics – especially in the 19th and 20 centuries, revealed the fundamental fact that the empirical world of physics and the formal world of mathematics – as they are accepted by the scientific community – contradict in many respects, our natural, self-evident interpretations, our intuitions.

The problem of intuitive cognitions became a fundamental aspect of the scientific endeavor. On one hand, the scientist needs intuition in his attempts to discover new strategies, new theoretical and experimental models. On the other hand, he should be aware that intuitions are not – as Descartes and Spinoza believed – the ultimate guarantee, the main foundation of objective truths. He, the scientist, has to learn to distinguish between information – an interpretation provided as rigorously as possible, by objective facts – and his subjectively imposed intuitions. It seems simple, but it is not. Such a distinction can never be absolute. Gödel's proof that in the realm of a mathematical system, apparently absolutely rigorous, one may find theorems, the truth of which is not decidable (provable) inside the respective system. Such an apparently closed system is, in fact, never absolutely closed. Heisenberg's principle of uncertainty implies that data of observations are never purely 'objective'. They are influenced by the observation process itself.

Cantor's 'paradise' such as Hilbert called it – the hierarchical world of transfinite numbers, so rigorously defined and built by Cantor, is defective in the core of its very structure. Russell discovered paradoxes hidden in the set theory. (See Kline's 'Mathematics: The loss of certainty', 1980).

Objective facts and intuitive interpretations are not absolutely distinguishable. We are witnessing a – probably – infinite process of this tendency towards absolute rigor and consistency. It is, in our opinion, this process which strongly emphasized, for the scientific community, the importance, the scientific relevance of the problem of intuitive cognitions as opposed and interacting with logically, objectively justified knowledge. Unfortunately cognitive psychology remained almost impermeable to these fascinating cognitive challenges.

A second source of the renewed interest for intuitive knowledge comes from the didactics of science and mathematics. When you have to teach a chapter in physics or mathematics, you usually discover that, what has already become clear to you – after long university studies – still encounters basic cognitive obstacles in the student's understanding. You, as a teacher, feel that the student – often ready to memorize what he is taught – does not, in fact, understand and memorize *genuinely* the respective knowledge. His intuitive grasp of a phenomenon is, very often, different from the scientific interpretation. This resistance may be tacit or open, implicit or explicit, but it exists. You teach the student that a body falls as an effect of the gravitational force. After some weeks, asking the same student to explain why an object falls, traveling to the ground, the answer is: 'Because the object is heavy'. 'Heaviness' is, for the student, a sufficient explanation. 'Heaviness' appears subjectively, intuitively to be an intrinsic property of the object, like density, or roughness. The idea that the earth exerts a pulling force on the object is not natural. Nothing, *in the direct experience of the student*, needs such an explanation and leads to it. The idea that the table on which the book is placed exerts a force upon the book (as if pushing it) is too fantastic for the naive imagination that it cannot be accepted, directly. And so on.

The same in mathematics. The idea that a square is a parallelogram is intuitively not less strange for many children. The idea that by multiplying two numbers, one may obtain a result which is *smaller* than one or both of the multiplied numbers, is also difficult to accept. The notion of an empty set seems sheer nonsense. The statement that the amount of points in a square and in a cube is the same, belongs to the same counter-intuitive notions.

As one advances in physics and in mathematics, one encounters concepts and statements more and more difficult to internalize and genuinely accept. Why? Because our natural interpretations of phenomena are adapted to the social and physical conditions of our terrestrial environment. The teacher discovers that the knowledge which he is supposed to transmit to the student collides, very often, with beliefs and interpretations which are naturally very resistant and conflicting with the scientifically established notions.

Consequently, the teacher, the author of textbooks, the didactical researcher become concerned with a psychological domain to a great extent ignored before: the intuitive interpretations, the beliefs of the student, with regard to the information he is asked to learn and assimilate. In comparing the role and the development of intuitive beliefs in the history of science and in the mind of the learner, one discovers profound analogies which may be helpful in explaining the respective facts.

A first step is then to identify, classify and describe the contribution – sometimes positive, sometimes negative – of intuitive structures in the history of science and mathematics, and in the didactical process. But there is a second step that should take us much further: to explain the mechanisms of shaping and functioning of intuitions. The mechanisms of intuitions are naturally hidden in our subconscious, mental activity. By their very nature, intuitions appear as sudden, global, synergetic reactions, as opposed to logically-based cognitions, which are, by definition, discursive and analytical.

Decoding the implicit mechanisms of intuitions is a very hard task. Introspection may be helpful, but it is not always trustworthy. Particular experimental conditions are usually needed. We are presently at the beginning stages of this fascinating enterprise.

A basic assumption, which will be described in the present paper is that intuitions constitute, generally, moments, states of transition from intellectual schemata (which are by their very nature, sequential processes) into adaptive reactions. (The sequential nature of schemata refers to schemata already build; not to the process of *producing* schemata which is not sequential but complex and hesitating.) The moment of transition conserves and expresses the acquisition stored in the corresponding schemata. On the other hand, it possesses the properties of suddenness, globality, extrapolativeness required by prompt, adequate behaviorally adapted reactions. Explaining an intuition would then require, first of all, the possibility to identify the schematic structure on which the respective intuition is based and, on the other hand, the means by which the sequential organization of the schema is converted in an apparently global, apparently self-evident cognition, leading to, and controlling the respective adaptive reaction.

### *The concepts of intuition and logical knowledge*

Let us consider some examples. Let us start with two arithmetical problems:

a) One litre of juice costs 5 shekels. How much will 3 litres of juice cost? How do you arrive at the answer?

The solution is simple and direct. If 1 litre costs 5 shekels, then three litres will cost 3 times 5. That is: 15 shekels.

b) One litre of juice costs 5 shekels. How much will 0.75 litre of juice cost? How do you arrive at the answer?

This problem is less simple than problem (a). Some students will answer that one has to divide $(5 : 0.75)$. Some will not answer at all. In this case, the correct answer is not direct. One has to think a little before proposing an answer. As a matter of fact the two problems have the same mathematical structure: the price per unit and the number of units. In both problems the operation is multiplication (the price per unit $\times$ the number of units). Nevertheless in problem (a) the correct answer is given without hesitation, directly as an obvious solution. In problem (b) the correct answer is not direct, it needs some logical reflection.

One says that the correct response in the first problem is obtained *intuitively*, while in problem (b) the correct response is obtained only indirectly by a *logical* endeavor. In problem (b) the direct, intuitive response is a mistaken one $(5 : 0.75)$ The correct response $(5 \times 0.75)$ contradicts the first intuition. The usual intuitive tendency is to claim that 'multiplication makes bigger' while 'division makes smaller'.

Let us consider a second example.

In Figure 1a there are two intersecting lines. They form two pairs of opposite angles. Let us refer to the pair of angles, $\alpha$ and $\beta$ and let us compare them. The direct answer is that the opposite angles $\alpha$ and $\beta$ formed by intersected lines are equal. Are we certain about this? Yes, the equality of the angles is self-evident. Let us generalize the question. How are the pairs of opposite angles formed by crossing lines? The direct answer is, usually, that *in general*, such pairs of angles are equal. That is, our answer $(\sphericalangle\alpha = \sphericalangle\beta)$ does not express a particular perception 'I *see* that the two angles are equal', *but*, rather, a general statement. As a matter of fact, as it has recently been shown (Stavy, Tirosh and Tsamir, 1997), the intuitive equality of two opposite angles formed by two crossing lines is not absolute. If on one side of the image the lines are longer (see Figure 1b), many subjects claim that the angles with longer 'arms' are bigger. In this case, the definition of the angle does not control any more the interpretation. The angles become integrated in the visual blocks (objects) consisting of the angles *and* their arms. Intuitions are not absolute. They depend on the context – in the present case, the perceptual context. Depending on the figure (1a and 1b), one has two different situations. One in which the intuition and the logically proven statement coincide (Figure 1a), and another (Figure 1b) in which a contradiction appears between the intuitive reaction and the proven statement.

In mathematical reasoning not everything which appears intuitively (directly) to be true is really, always true. *One has to prove the respective*

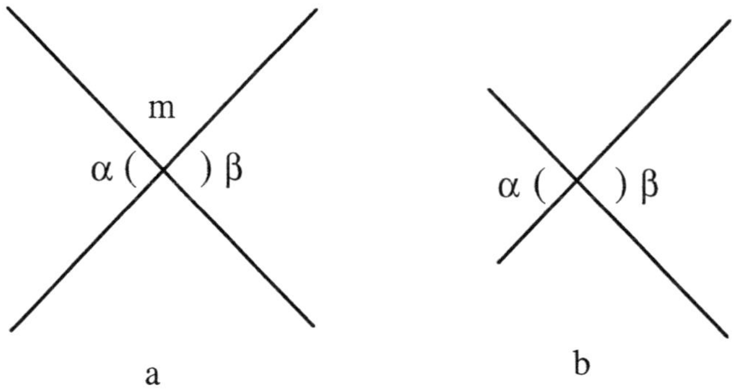


*Figure 1.*

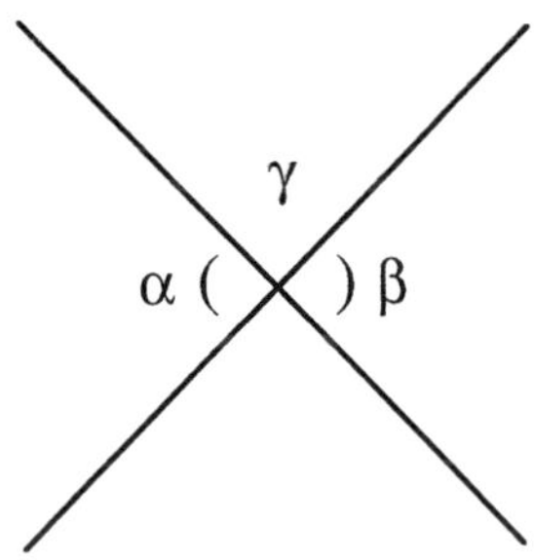


*Figure 2.*

*statement*. To an individual not trained in mathematics, such a claim may appear strange, but not to a mathematician. As a matter of fact the above statement may be proved. Let us consider, again, the two intersecting lines (see Figure 2). Let us *prove* that $\sphericalangle\alpha = \sphericalangle\beta$. One may write that $\sphericalangle\alpha + \sphericalangle\gamma = 180\,^\circ$; $\sphericalangle\beta + \sphericalangle\gamma = 180\,^\circ$ conclusion: $\sphericalangle\alpha = \sphericalangle\beta$.

Now the equality of the angles $\alpha$ and $\beta$ has been established *indirectly* by logical reasoning.

Let us consider the following statement:
The expression $n(n^2 - 1)$ is divisible by 6 for every natural number n. Is this statement intuitively evident? Usually, not. One has to prove it. First step: $n^2 - 1 = (n - 1)(n + 1)$. Consequently, one has: $n(n - 1)(n + 1) = (n - 1)\,n(n + 1)$, which represents three consecutive numbers, for instance 3, 4, 5. One remarks, first, that given three consecutive numbers, at least one should be an even number (divisible by 2), and one should be divisible by three. The product is, then, divisible by $2 \times 3 = 6$.

In the above example, one deals with a statement which does not appear intuitively (immediately) to be true. Only indirectly, through a formal proof, one may confirm its validity.

Let us consider a further example in the same sense. The statement: The sum of the angles in a triangle is 180°. Does the statement appear to be *intuitively* true? No, it does not. There are an infinity of various possible triangles. It is surprising to learn that *always* the sum of their angles equals two right angles. The above statement is *not* an intuitive cognition. It has to be proved, logically, indirectly, resorting to a number of successive steps. *Intuitively*, 180° is not more acceptable than, for instance 160°.

From the above examples one may learn that there are, in mathematics (and in science in general), statements which appear to be acceptable directly, as self-evident, while for other statements, a proof is necessary in order to accept them as true. One may generalize and hypothesize that intellectual cognitions present two basic forms:

a) A category of cognitions which appear directly acceptable as self-evident. These are intuitive cognitions.
b) A category of cognitions which are accepted indirectly on the basis of a certain explicit, logical proof. These are logical, or logically-based cognitions.

An important remark: Not every direct cognition is an intuition. Perceptions are directly grasped by senses, but they are not intuitions. Intuitions are *intellectual* cognitions – expressing a general conception (a notion, a principle, an interpretation, a prediction, a solution) while perceptions are sensorial cognitions (for instance: I *see* a chair, a triangle, etc.).

*Various types of relationships between intuitions and formally-based cognitions*

One may distinguish the following situations referring to the relationships between intuitive and formal knowledge:

a) Statements which are accepted without proof, only on the basis of their intuitive evidence. In the Euclidean geometry, such statements provide a number of axioms.
   For instance:

   - Two points determine a straight line and only one.
   - The shortest distance between two points is a straight line.
   - Through a point outside a straight line, one may draw a parallel and only one to that line (the famous 5th postulate of Euclid).

   As one knows, in modern mathematics, axioms are no longer established by their apparent self-evidence, but in accordance with the axiomatic system established by the designer of the system.

b) Statements which appear *intuitively* to be true but, despite this, may be and should be proved. For instance:

- The opposite angles formed by two intersecting lines are equal (the case in which the image is symmetric).
- In a triangle, the sum of two sides is always bigger than the third side.
- In an isosceles triangle (two equal sides) the angles adjacent to the basis are also equal.

c) Statements which are not self-evident and which have to be proved in order to be accepted. For instance:

- In a triangle the sum of the angles is equal to two right angles.
- In a right-angled triangle, the sum of the squares of the cathetae is equal to the square of the hypotenuse $a^2 + b^b = c^2$ (the Pythagoras theorem).
- In a triangle, a straight line parallel to one of the sides divides the two other ones in proportional line segments (the theorem of Thales).

Again, in formal, modern mathematics, there is no longer the absolute distinction which existed in the Euclidean view between axioms, theorems, definitions and laws. The notion of 'law' (e.g., the commutative law) can be an observed 'fact' for children, an *axiom* in, say, the notions of group, field, ring and a *theorem* in the context of Paeon's postulates.

d) The most interesting situation is that in which a conflict appears between the intuitive reaction to a given situation and the cognition reached through a logical analysis.

Let us come back to a problem mentioned above: 'One litre of juice costs 5 shekels. How much will 0.75 litres cost? By which operation would one solve the problem?'

The first tendency of many people is to answer: 'By division' – considering that 'division makes smaller'. In order to get the correct answer one may resort to the proportional relationship between cost and quantity: the price is proportional with the quantity:

$$\frac{quantity(1)}{quantity(2)} = \frac{price(1)}{price(2)}$$

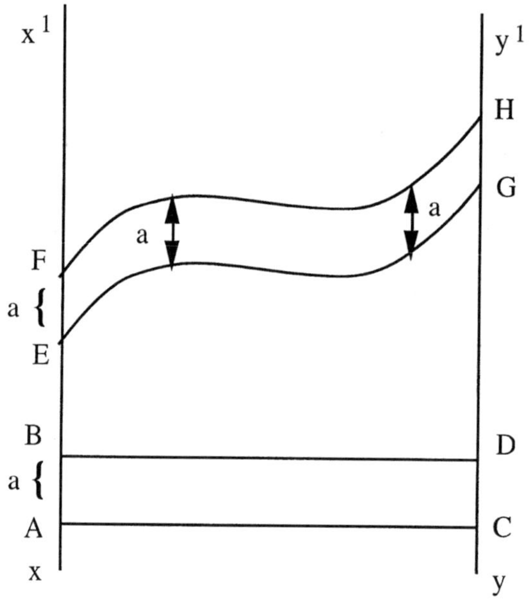


*Figure 3.*

And then one has:

$$\frac{1}{0.75} = \frac{5}{x}$$

$$x = 5 \times 0.75$$

The operation is – surprisingly – a multiplication. The problem may also be solved by analogy referring to a problem with the same mathematical structure but using whole numbers.

In this example a conflict appears between the first intuitive solution and the correct logical solution.

Let us mention a second example. Let us consider two parallel axes $xx^1$ and $yy^1$ (see Figure 3).

Let us choose a certain distance $a$ on $xx^1$. From A and B let us draw two parallel lines perpendicular to $xx^1$ and $yy^1$.

Let us now consider again, two points on $xx^1$ at distance $a$. Starting from E and F, let us draw this time, two curves so that the distance between the two corresponding points remains the same, that is $a$.

One asks to prove that the two areas ACDB and EGHF are equivalent. We posed this problem to high school students. The immediate reaction of these students was that the two areas are not equal and this means that there is nothing to prove. This was the *intuitive*, direct, apparently self-evident reaction to our problem. In reality, the two areas *are* equivalent, but this may be proven only indirectly by a logical analysis. One may find a solution through integrals, but there is also a more elementary procedure.

Let us consider the following areas: area ACGE and area BDHF. These two areas are equivalent because the second may be obtained from the first by translation (distance $a$). From these two equivalent areas one has to subtract the common part, which is area BDGE. What remains are the two initial areas ACDB and ECHF. If, from two equal quantities, one subtracts the same quantity, the quantities which remain are equal. The equivalence of the considered areas is surprising. *Intuitively*, the two areas appear non-equivalent, but the logical analysis proves that they *are* equivalent.

e) One may also describe situations in which two different representations of the same problem may lead to contradictory intuitions. Considering the set $\{1, 2, 3, 4, 5, 6 \ldots\}$ it seems *intuitively* that the set of natural numbers and the set of even numbers are *not* equivalent. But let us consider the following representation:
$\{1, 2, 3, 4, 5, 6, \ldots\}$
$\{2, 4, 6, 8, 10, 12 \ldots\}$

In this representation, to *each* natural number corresponds an even number. The two sets are intuitively equivalent.

Summarizing one gets the following:

- A situation in which a statement is accepted intuitively and no prove is requested.
- A situation in which a statement is accepted intuitively, but in mathematics it is also formally proved (coincidence between intuitive acceptance and a logically-based conclusion).
- A situation in which a statement is not intuitive, self-evident, and may be accepted only on the basis of a formal proof.
- A situation in which a conflict appears between the intuitive interpretation (solution) regarding a statement and the formally-based response.
- A situation in which two conflicting intuitions may appear.

The teacher has to be aware of these possible situations in order to understand and solve the difficulties of the students in learning mathematics.

*Didactical implications*

a) It would seem, at first glance, that the most favorable situation for the teaching process is that in which the intuitive solution coincides with that obtained via logical analysis. But the didactical reality is different.

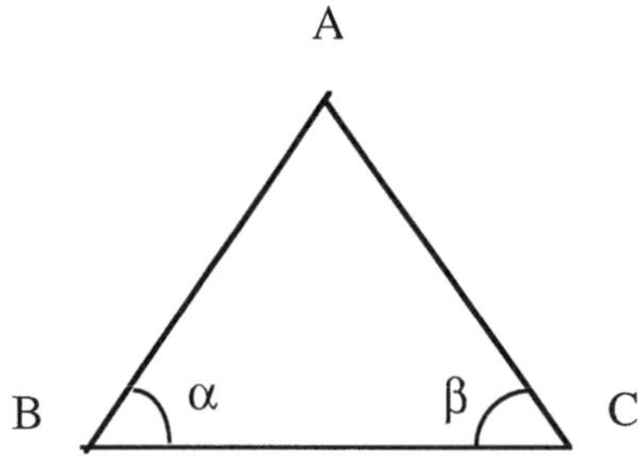


*Figure 4.*

Let us remember the example of the isosceles triangles: Knowing that AB = AC, one has to prove that $\sphericalangle\alpha = \sphericalangle\beta$ (see Figure 4). The student's immediate reaction is: 'What for'? 'Why shall we prove what is evident'? The student does not accept that an intuitively evident statement should be formally proven. The proof seems to be superfluous to the student and the requirement to prove a statement which appears obvious may strengthen the student's feeling that mathematics is an arbitrary, useless, whimsical game. We may add other examples in the same sense.

Let us remember the properties of commutativity and associativity of the operations of addition and multiplication. Starting from a certain age, the student would think: $a + b$ and $b + a$ lead to the same result; the same for $ab = ba$. Why should one emphasize such trivial, evident facts as basic mathematical properties? As above, *the intuitiveness of a certain property tends to obscure in the student's mind the mathematical importance of it.* An apparently trivial property seems to discard the necessity and utility of mentioning it explicitly, of proving it or defining it.

The same for the law of associativity in regard to addition and multiplication. If one considers, lets say, three numbers a, b, c, the sum will be the same, no matter whether one adds first $a + b$ and then, c, or whether one adds a with the sum of $b + c$. Using parentheses, one writes: $(a + b) + c = a + (b + c)$. Similarly: $(a \times b) \times c = a \times (b \times c)$. For that reason, one does not use parentheses when adding or multiplying whole numbers.

All this seems to be intuitively evident, and then, mentioning the laws of commutativity and associativity with regard to whole numbers, seems to be unjustified, superfluous.

The reasons for describing as laws, these *apparently* trivial facts are the following:

- In mathematics, *every* property, every accepted statement or notion should be identified explicitly as being an axiom, a theorem, a definition, a law (a general property), a basic or derived notion. Thus, for the fundamental reason that mathematics is a deductive, formal, rigorous system, intuitive evidence does not represent, *by itself*, in mathematics an accepted justification.

- A second reason, no less important, for which such apparently trivial properties should be considered and emphasized explicitly, is that they are not automatically applicable to every mathematical operation. The laws of commutativity and associativity do *not* hold for subtraction and division:

$a - b$ is not equal $b - a$; $5:3$ is different from $3:5$; $12:(4:2)$ is different from $(12:4):2$; $12:(4:2) = 6$; $(12:4):2 = 1.5$.

Let us consider a second example, namely the relation of equivalence. Why should one mention that the relation of equivalence, $A \equiv B$ is defined by three properties? If $A \equiv B$, then $B \equiv A$ (symmetry) and $A \equiv A$ (reflexivity). From $A \equiv B$ and $B \equiv C$ it follows that $A \equiv C$ (transitivity).

The following numerical examples seem to emphasize the triviality of the intuitive evidence of these properties.

$$12 : 2 = 6 \rightarrow 6 = 12 : 2;$$
$$12 = 12;$$
$$12 : 2 = 6; 6 = 2 \times 3; \rightarrow 12 : 2 = 2 \times 3.$$

All this is true, but trivial, evident (intuitively). Why are these facts worth being emphasized? Because things are not always so. Not every mathematical relation has these properties.

First, the concept of equivalence is more general and much more subtle than the concept of equality. What seems to be trivial for equality, is not necessarily trivial for equivalence. The first thing to mention is that there are various relations for which the above properties do not hold. Let us consider, for example, the relation of order: $A > B$. This relation is *not* symmetric, *not* reflexive, but transitive.

From $A > B$, it does not follow that $B > A$. On the contrary: $B < A$ (the relation is anti-symmetric). From $A > B$ certainly it does not follow that $A > A$. On the contrary. But from $A > B$, $B > C$, *it follows* that $A > C$. This relation *is* transitive.

Let us consider another example. Dan is asked to solve the following equation:

$$\frac{x}{2} + \frac{x}{3} = 5$$

He has learnt that, first, he has to find and eliminate the common denominator. He then writes:

1. $^{3)}\frac{x}{2} + {}^{2)}\frac{x}{3} = {}^{6)}5$

2. $3x + 2x = 30$
3. $5x = 30$
4. $x = 6$

What Dan does not, probably, know is that the three equations are equivalent. By definition two equations are equivalent if they have the same truth set, the same roots. Dan has transformed the equations, but in such a way so as to preserve the same truth set. By eliminating the common denominator, he in fact, has multiplied both sides of the equation with the same number. He, thus obtained a new equation equivalent to the former one. He has also applied the transitivity property: Equation (1) and equation (4) are equivalent. As far as we know, the concept of equivalence in general and the concept of equivalence of equations are not taught explicitly at the school level.

The above examples refer to the situation mentioned above: When a situation (for instance, a property, a statement) seems to be trivial at first glance, the student has the feeling that proofs, definitions are superfluous. The reason for this situation is that the student has not fully grasped the meaning of mathematics as a deductive, formal, rigorous body of knowledge. He tends to interpret mathematical facts in an empirical way, that is, to refer to mathematical objects and operations as he does when referring to concrete realities. No-one has asked himself if a chair is a chair, if changing the location of a chair, the chair remains the same, etc.; if sitting on a chair changes its function, shape, color, etc.

*In mathematics, intuitive evidence of a property or operation does not exclude the necessity to confer on them a formal, rigorous status (definition, proof, etc.) in accordance with the axiomatic, deductive structure of mathematics.*

b) In the above lines we have considered a situation in which intuitive evidence seems to eliminate the need for a formal description or justification.

Another, very important, situation occurring often in mathematics teaching (as already mentioned) is that in which intuitive evidence *conflicts* with the formal status (determined by a definition, a theorem, a formal proof, a formal property). In this case, the teacher has to identify the intuitive tendencies of the student and to try to explain their sources. A main didactical procedure for helping the student to overcome the difficulty is to make the student aware of the conflict and to help him to grasp the fundamental fact that *in mathematics what finally decides* is the formal status of the respective entity.

Let us consider again, some further examples:

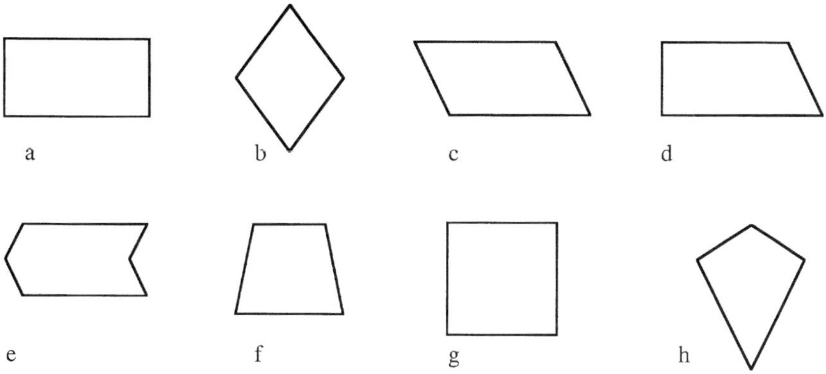


*Figure 5.*

A number of students in grades 9, 10 and 11 were asked to define the term 'parallelogram' and were presented with the following figures (see Figure 5).

Correct definitions of the parallelogram were given by 88% (gr. 9), 90% (gr. 10), and 88% (gr. 11).

In grade 9, 75% gave correct definitions of the term parallelogram and also identified among the figures, those which were parallelograms. In grade 10, 67% answered this way. In grade 11, only 61% of the students defined correctly and identified correctly the parallelograms among the above figures (that is, according to the definition).

Considering separately the figures presented, one finds that *all* the students of all age levels identified correctly Figure c, while Figures a and g, for instance, were identified only by 85% as parallelograms. In grade 11, only 76% of the students identified the rhombus as a parallelogram.

In this example, a conflict appears in some students between the intuitive interpretation of the figure, *according to its visual structure*, and the formal definition. In *younger students*, the conflict is still stronger. What happens in reality is that it is intuitively difficult to include in the same category figures so different in shape such as a square and an oblique parallelogram. Their visual structure is different. On the other hand, these two figures may be estimated to belong to the same category when considering a sufficiently broad definition (like: a parallelogram is a quadrilateral if its opposite sides are parallel (or equal). Only when the student is sufficiently trained in mathematics, is he able to rely *not on the figural aspect*, but on the formal definition (see Fischbein and Nachlieli, 1997).

Let us consider some other examples in which the intuitive impression conflicts with the formal (mathematical solution). In a research, the following question was posed:

'In a game of Lotto, one has to chose 6 numbers from a total of 40. Vered has chosen 1, 2, 3, 4, 5, 6. Ruth has chosen 39, 1, 17, 33, 8, 27. Who has a greater chance of winning?

- Vered has a greater chance of winning.
- Ruth has a greater chance of winning.
- Vered and Ruth have the same chance of winning.

The correct answer, of course, is that the chances are equal, because every group of six numbers has the same probability. Nevertheless, many students considered that Ruth has a greater chance of winning.

*Intuitively*, it seems practically impossible that an ordered sequence of numbers may win. A random sequence seems to better represent the randomness of the lotto game. This is called the *representativeness* bias. According to age (grades) the following percentages of mistaken answers was obtained: 70% (gr. 5), 55% (gr. 7), 35% (gr. 9) and 22% (college students). One may see from the above data that the reactions improve with age (Fischbein and Schnarch, 1997).

A further example is taken from the domain of actual infinity. Students (grades 5–9) were asked to compare the number of elements in the set of natural numbers with that of the set of even numbers: one knows that the two sets are equivalent. But most of the students, at various age levels answered, intuitively that the set of natural numbers is 'bigger'. This is the intuitive answer which conflicts with the mathematical, formal one. This answer was given by 60.9% (gr. 5), 77.6% (gr. 6), 63.2% (gr. 7), 80.8% (gr. 8) and 73.4% (gr. 9) of subjects (Fischbein, Tirosh and Hess, 1979, p. 18). As we have seen above, the conflict disappears if the numbers of the two sets are arranged in a corresponding way:

$$\{1, 2, 3, 4, 5, 6, \ldots\}$$

$$\{2, 4, 6, 8, 10, 12\ldots\}$$

The above examples refer to situations where there is a conflict between an intuitive (direct, self-evident) cognition and a formal, mathematical truth based on formal constraints (definitions and deductively-based theorems). Such situations appear very often in the teaching of mathematics. The student may not be aware of the conflict. The effect is that the student does not accept, does not understand the formal statement, or even when he seems to understand initially, he tends to forget it and the intuitive interpretation is that which decides the student's solution. *In our opinion it is helpful that the students should become aware of the conflict.* Simply ignoring the conflict (that is, the intuitively erroneous reaction) leaves untouched the original intuition. Thus the conflict remains latent and finally the student will, probably, forget the formal, mathematically correct answer.

c) A third category of situations is that in which formal, mathematical statements are simply not related to any intuitive representation. The student has to rely only on the formal mathematical truth, on formal proofs and definitions.

- Many mathematical statements and formulas are in that situation. For instance, the formula for solving quadratic equations is in that situation. It does not rely on any intuitive support.

Let us mention some more examples:

- The equality $a^0 = 1$ is fixed by definition. The expression $a^0$ has no direct, unique, intuitive meaning. We have found that some students answer $a^0 = 0$, others write $a^0 = 1$, or $a^0 = a$, while others do not answer at all. The definition $a^0 = 1$ should be accepted in order to be consistent with the mathematical constraints. If one has accepted that $\frac{a^m}{a^n} = a^{m-n}$, then one has, on the one hand $\frac{a^m}{a^m} = a^{m-n} = a^0$, and on the other hand $\frac{a^m}{a^m} = 1$. That is: $a^0 = 1$.
- A further example refers to operations with negative numbers. We know that $(-a) \times (-b) = +ab$. There is no direct intuitive justification for such a rule. Intuitively $(-a) \times (-a)$ has no meaning. Nevertheless, in order to conserve the consistency of mathematics, one has to accept that $(-a) \times (-b) = +ab$. Several attempts have been made to produce models justifying that rule intuitively, but, generally, they were too complicated and therefore they were of no didactical utility.

- Imaginary numbers have no intuitive meaning. *Intuitively* $\sqrt{-1}$ does not reveal any intuitive association.

*For counter-intuitive notions, a new intuitive context may be created*, for instance, as in the case for complex numbers, but the intellectual process to produce, understand and use them, is so difficult that it is practically impossible to transfer their 'intuitiveness' to the original symbols. The original symbols (for instance, $\sqrt{-1}$), are still counter-intuitive.

The student has to get used to the idea that mathematics is, by its very nature, an abstract, formal, deductive system of knowledge. *Intuitive models are very often useful, but they are not always possible.* It is not recommendable to strive to invent artificial intuitive models for every concept or operation. It would be an absurdity to try to invent intuitive models, for instance, for $a^0$ or $a^{-n}$.

d) A fourth possible situation is that in which one deals with mathematical concepts or operations which are both difficult to accept intuitively (counter-intuitive) and difficult to handle formally. There are the situations which are didactically, the most difficult.

Let us consider an example. Let us refer to the operation of subtraction.

$$\text{Case A}: \quad \begin{array}{r} 7635 \\ -\,\underline{5421} \\ 2214 \end{array}$$

Such a subtraction does not pose any problem. According to the rule, one has to subtract every digit of the second number from every corresponding digit of the first number. Intuitively and formally, this operation is evident and simple. But usually, the operation of subtraction is not so simple.

$$\text{Case B}: \quad \begin{array}{r} 1702 \\ -\,\underline{1368} \end{array}$$

According to the above rule, start subtracting from the right. First one has to subtract 8 from 2. *Intuitively it does not work.* It has been found that, sometimes, children reverse the operation (8–2) and write 6. This first step is counter-intuitive. What one has to do is to borrow from the next digit to the left. But this is also not possible, intuitively. The next digit is 0. So you have to borrow again from the next digit to the left which is 7. Now it works, but in the meantime the student has, possibly forgotten what he is doing. In order not to lose his way, the student has to understand the principle of the place value of the digits, which is far from being simple. It expresses a formal convention – the value expressed by a digit depends on its place in the number. The above combination of intuitive and formal-procedural difficulties explains why the operation of subtraction appears to be so difficult for many children.

*Generally speaking, for the teaching of mathematics, it is very important that the teacher understands the interactions between the intuitive, the formal and the procedural aspects in the processes of understanding, remembering and problem solving.* If the intuitive forces are neglected, they will, nevertheless, continue to influence the pupil's capacity of understanding and solving, but unfortunately in an uncontrolled manner, usually disturbing the mathematical thinking process. If the formal aspect is neglected and one would tend to rely exclusively on intuitive arguments, that which will be taught will not be mathematics.

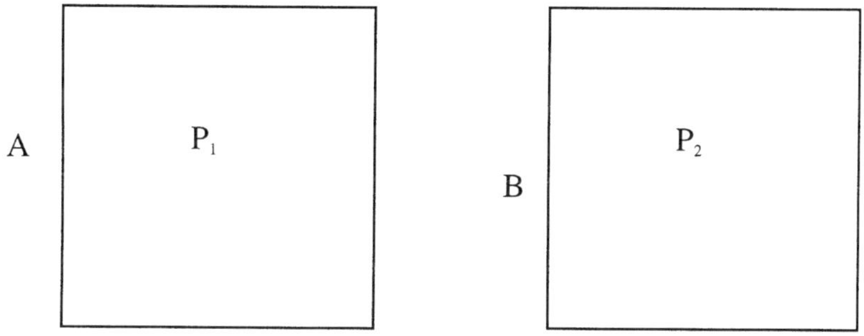


*Figure 6.*

*The general characteristics of intuitive cognitions*

So far we have talked about intuition as a direct, self-evident kind of knowledge. This is the basic, most salient property of an intuitive cognition. But a more complete description of intuitive cognitions is necessary in order to understand their role in a reasoning process.

- *Direct, self-evident* cognitions mean that intuitions are cognitions accepted as such without the individual feeling the need for further checking and proof. The statement 'The shortest distance between two points is a straight line' is such a self-evident, directly acceptable statement.
- *Intrinsic certainty*. An intuitive cognition is usually associated with a feeling of certitude, of intrinsic conviction. The above statement concerning the straight line is, *subjectively*, felt *as certain*. 'Intrinsic' means that no external support is required for getting this kind of direct conviction (formal or empirical).
- *Coerciveness*. Intuitions exert a coercive effect on the individual's reasoning strategies and on his selection of hypotheses and solutions. This means that the individual tends to reject alternative interpretations, those which would contradict its intuitions. We have mentioned above that, usually, pupils and even adults believe that 'multiplication makes bigger' and 'division makes smaller'. They have gotten used to this belief from their childhood when they operated only with natural numbers (for which these beliefs are correct). Later on, even after learning the notion of rational numbers (that is, including also sub-unitary fractions) they continue to hold the same belief – which obviously, does not correspond any more.

- *Extrapolativeness.* An important property of intuitive cognitions is their capacity to extrapolate beyond any empirical support. For instance:

  - The statement 'Through a point external to a straight line, one may draw one and only one parallel to that line', expresses the extrapolative capacity of intuition. No empirical or formal proof can support such a claim: one considers infinite lines. Nevertheless we accept, intuitively, the respective statement, as certain, as self-evident, *extrapolating* from what we are practically able to achieve (two straight lines which do not meet, though contained in the same plane). *We feel* that we may go on *indefinitely* this way. *The extrapolation comes from the intuitive cognition itself.* The extrapolative capacity is in the nature of intuition. The strong intuitive character of this Euclidean axiom has prevented mathematicians from replacing it with alternative non-intuitive axioms until the 19th century, thus building different, non-Euclidean geometries.
  - In the same vein, one may remember the statement: 'Every (whole) number has a successor'. This is not a statement for which we feel the need of an empirical or formal proof. We simply, naturally, *extrapolate* beyond any sequence of numbers. This extrapolative capacity intervenes also in the principle of mathematical induction on which the method of proving called: 'mathematical induction' is based. This principle states the following: 'In order to prove that a statement P(n) is true, one resorts to two steps: a) One has to prove that the statement P(n) is true for $n = 1$, and b) One has to prove that if the statement P(n) is true for $n = k$, it is also true for $n = k + 1$ (n is any natural number). Consequently, if one has proved that the statement is true for $n = 1$, it follows that it is also true for $n = 2$, and then for $n = 3$ and so on for every natural number. The statement is then true for every n (by extrapolation). In this type of proof, *one relies on the extrapolative capacity of intuition.*

- *Globality.* Intuitions are global cognitions in opposition to logically acquired cognitions which are sequential, analytical. Let us consider some examples:

Let us consider the following:
A 4–5 year-old child is presented with two sheets of paper. On sheet A (see Figure 6), the experimenter draws a point ($P_1$) and asks the child to draw a point on sheet B 'on exactly the same place' as the point $P_1$ on sheet A (see Figure 6).

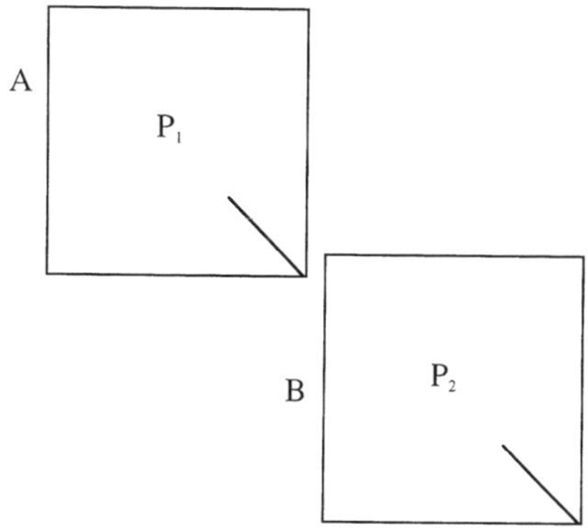


Figure 7a

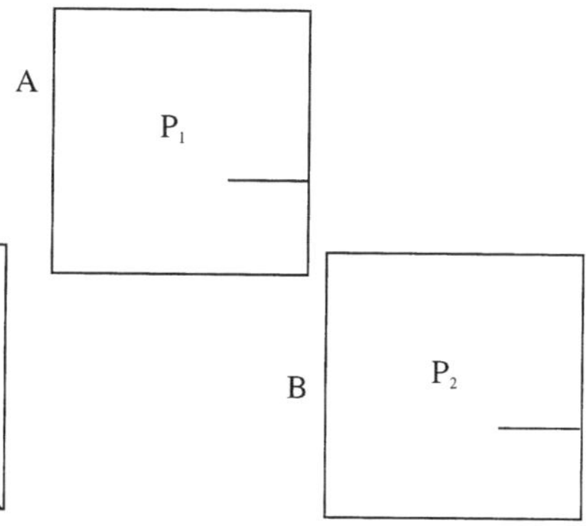


Figure 7b

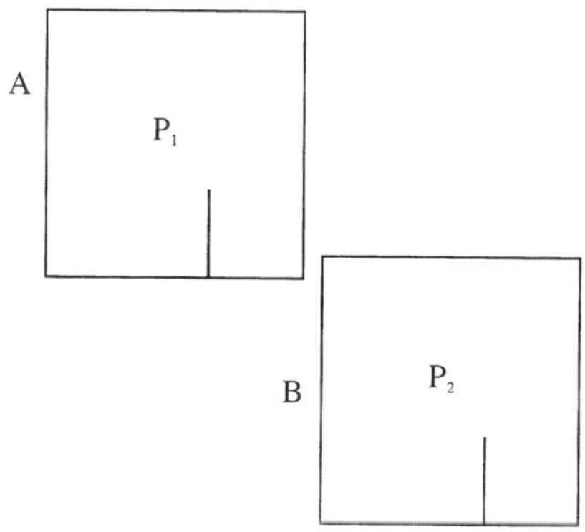


Figure 7c

*Figure 7.*

The child will usually draw a point ($P_2$) on sheet B more or less with the same location. If he is asked to explain why he puts the point in the respective place ('Why did you put the point here?'), the child will not be able to give an explanation. He solved the problem *intuitively*, directly, through a *global* estimation. The location was not determined through an act of measuring, which is *an explicit, logical, analytical act.*

If one asks older children, say aged 6–7, the same question, some of them will try a one-dimensional justification (see Figures 7a–c). The children explain: 'I looked at the first sheet to see how far the point is from the line (the edge)'.

In this case the reaction is a mixture of an intuitive and a logical-analytical solution. *We witness the very interesting phenomenon of a first step of a process by which a logically-based solution emerges from a primitive global estimation.*

Some of the older children (aged 10 and over) were able to locate the points using two coordinates – usually the distances from two perpendicular edges of the sheet. As a matter of fact, only about 17% – 12-year-old children reached spontaneously this analytical solution. The others had to be taught the full formal technique, but they are already prepared at this age to understand the principle of two coordinates necessary for locating a point in a two-dimensional space.

With the above example, we have tried to illustrate the notion of a *global* solution: *an intuitive (direct) response is a global response as opposed to a logical – analytically-based solution.* (For more details regarding the acquisition of the notion of coordinates in children, see Piaget, Inhelder and Szeminska, 1960, pp. 153–172; and Fischbein, 1963, pp. 401–414.)

Let us mention another example with regard to the global character of intuitive estimations:
A 4–5 year-old child is presented with two rows of marbles (row A and row B). Let us consider the following situations:

a) Row A and row B are of the same length. Asked to compare the number of marbles in the two rows, the child affirms that there are the same number of marbles in the two rows even if the numbers are different.
b) Row B is longer – the child will conclude that in row B there are more marbles, even if in the two rows it is the same number of marbles. The child does not count the marbles. His response is given as a *global estimation, directly*. Piaget claims that at this age the child thinks through *configurations*. In the present case, one deals with *intuitive estimations* in which the main visual impact, the length of the rows, plays a decisive role.

If the same question is posed to a 6–7 year-old child, he will, usually, count the marbles before responding. This is a completely different attitude. His answer is based on a logical, analytical operation, i.e., the process of counting.

All the examples mentioned previously, referring to intuitive cognitions are also examples of global, direct estimations.

Let us add another example. Let us ask students – even college students – to represent the fraction $\frac{1}{3}$ by a decimal number. The students' answer is, usually $\frac{1}{3} = 0.3333\ldots$ Let us ask the same question in a different manner: 'The decimal number 0.3333 … is equal to $\frac{1}{3}$ or *tends* to $\frac{1}{3}$?' The students

– even college students – respond, usually that 0.3333 … *tends* to $\frac{1}{3}$, but never reaches it because one deals with an infinite number of digits and infinity can never be reached.

As a matter of fact, 0.3333 … $= \frac{1}{3}$. We know that equality is a symmetrical relation: If A = B, then B = A. If $\frac{1}{3} = 0.3333$ …, then it follows *necessarily* that 0.3333 … $= \frac{1}{3}$. The equality $\frac{1}{3} = 0.3333$ … is easily accepted intuitively because one deals here with a *potential infinity*, that is, a process which continues indefinitely. A 12 year-old child understands *intuitively* that a straight line may be extended indefinitely.

With *actual infinity* things are different. *Actual infinity* means a quantity (for instance, an infinite set of points, of numbers) which is given and should be grasped entirely as such. The set of points of a line segment constitute an actual infinity. Research has shown that, while a potential infinity may be *understood, grasped, accepted intuitively* – as an unlimited *process* – an actual infinity cannot be grasped intuitively, as a *given* quantity. For that reason, *problems including operations with actual infinities lead to deep intuitive difficulties* (see Fischbein, Tirosh and Hess, 1979). It is difficult to accept, for instance, that the set of natural numbers is equivalent to a subset of it, for instance, the set of even numbers. The difficulty of which we are speaking about with regard to the intuition of actual infinity may be explained in the following way: Intuition, as we have seen, means a *global*, synthetic grasp and interpretation of a situation. But our mind is, naturally, not prepared to grasp *in one view* an infinity of elements. In everyday life situations, we are used to dealing with finite realities (or with *processes* which go on). Consequently, when we are asked whether 0.3333 … is equal to $\frac{1}{3}$ or *tends* to $\frac{1}{3}$, the usual answer is that 0.3333 … tends to $\frac{1}{3}$. In order to accept, intuitively, that 0.3333 … *is equal* to $\frac{1}{3}$, one has to be able to grasp *globally*, directly the infinite multitude of the respective digits. As I said, we are, naturally, not adapted for such a global perception.

Let us summarize: The main characteristics of intuitive cognitions are: *Immediacy, self-evidence, intrinsic certitude, subjective coerciveness and globality.*

### *Affirmatory and anticipatory intuitions*

*All* the examples mentioned so far belong to the category of intuitions called by us *affirmatory intuitions*. These are statements, representations, interpretations, solutions which appear *to the individual* to be directly acceptable, self-evident, global and intrinsically necessary.

But one should distinguish a second category of intuitions namely, *anticipatory intuitions* which have not been mentioned so far. When striving to solve a problem, the solution of which is not direct, the search effort passes through a number of stages: a) First, one tries to grasp meaningfully, the question addressed, making use of the information displayed in the text of the problem. The solver has to understand clearly and distinguish *what is given* and *what is required*. b) Secondly, the solver has to mobilize various, previously acquired informations and associations, in such a manner as to bridge the gap between what is given and what is required. Such mental effort is sometimes tacit, sometimes conscious, explicit. c) When this endeavor comes to a well-structured end, the solver feels that he has reached the solution.

In reality, things are more complex. The solving process (see stage b) passes, usually, through three main phases. In the first phase (b 1) the solver invests his maximal efforts trying various strategies, hesitating, resorting to previously acquired solving schemata and models, rejecting inadequate solutions. This process is more or less conscious because the strategies used are relatively conscious. Very often the solver gives up turning to another activity or to a rest period (b 2). *Suddenly*, he has the feeling that he has found the solution. He does not possess, yet, all the elements of the solution, that is, the formal, analytical, deductively justified steps of the solution. What he has in mind, during the first moment, is *a global idea, a global representation of the main direction leading to the solution*. This is also an intuition, an *anticipatory intuition*, called, sometimes, the 'illumination' moment. What characterizes such an intuition is, first of all, the fact that it represents *a moment in a solving endeavor*. Secondly, such an intuition is associated with a feeling of deep conviction, a feeling of certitude, *before* the entire chain of the formal – analytical basis of the solution has been established by the solver. For a mathematician, the solving process is not concluded before he is able to invoke explicitly *all* the arguments supporting the initially guessed solution.

Briefly speaking, what characterizes anticipatory intuitions, are the following aspects: a) They appear during a solving endeavor, usually, suddenly after a phase of intensive search. b) They present a global character. c) In contrast to a usual guess or hypothesis, these intuitions are associated with a feeling of certitude, though the detailed justification or proof is *yet* to be found.

The well known American psychologist Jerome S. Bruner has claimed that in the process of intellectual education, the students should be encouraged to express such intuitions during classroom activities (Bruner, 1965). Many teachers used to react negatively when a student suggests a solution or a solving strategy before he is able to support his response by a complete, well-organized justification. The effect is that, usually, students do not dare to express their views during a collective solving endeavor in the classroom. The capacity to estimate the plausibility of a solution strategy should be educated and this cannot be done if the student does not have the opportunity to confront his anticipatory intuitions with the opinions of his colleagues.

*Correctness of intuitions*

Let us emphasize that the difference between correct and incorrect intuitions is only a relative one. It depends on the conceptual setting in which the intuitions are considered. Let us mention some examples:

a) People used to claim that 'multiplication makes bigger' and 'division makes smaller'. It is an intuitive claim based on the early experience of the child with natural numbers. As long as one considers the set of natural numbers, the above sentences and the respective intuitions are correct. But when a college student makes the same affirmation, one has an incorrect intuition because one should assume that the student learnt long before about fractions smaller than 1, for which the above claims (intuitions) are no longer correct.

b) Considering the setting of our earthly experience, a launched object travels as long as a force supports the motion. This is the primary intuition one usually has with regard to the motion of objects. This is a correct intuition as long as one refers to the motion of objects in the vicinity of the earth. A force is always needed on earth because one has to consider the friction of the air, and the effect of gravitation. But, in the theoretical conditions of a space in which no gravitation and no air friction are assumed, the motion of the object will continue forever in a straight direction and with a constant velocity.

*Usually, the distinction between incorrect and correct intuitions imply the idea of comparing intuitions based on the limited, personal experience of the individual and the objectively proven conceptions accepted by the scientific community.*

*The effect of presentation of a problem on the intuitive reactions*

When analyzing an intuitive interpretation, one has to take into account the way in which the problem is presented. Let us remember an already mentioned example: The comparison between the set of natural numbers and the set of even numbers. The two infinite sets are mathematically equivalent because a one-to-one correspondence may be established:

$$M_1 = \{1, 2, 3, 4, 5, 6, 7 \ldots\}$$
$$M_2 = \{2, 4, 6, 8, 10, 12 \ldots\}$$

To each natural number corresponds an even number and vice-versa. This is visible, it is intuitively evident. But let us consider the same question posed in the following way:

$$M = \{1, \boxed{2}, 3, \boxed{4}, 5, \boxed{6}, 7, \boxed{8} \ldots\}$$

In this presentation one emphasizes the fact that the set of even numbers is a *subset* of the set of natural numbers. With that presentation a principle – a schema – *adequate only for finite sets* is triggered. It is intuitively very difficult to accept that a set of elements may be equivalent to a subset of it. It seems intuitively evident that the whole contains more elements than a part of it – this is true *but only for finite sets.*

Let us consider an example of a different type. Subjects were asked whether the set: $M_2 = \{a,b, \; a,b, \; a,b, \; a,b \ldots\}$ is finite or infinite. The correct answer contrary to the appearance, is that the set is finite and contains just two elements: a, b. In conformity with the properties of sets each element should be counted only once (Balzan, 1997).

Intuitions are very sensitive to the influences of the context, especially because no logical, formal support intervenes. Such examples are very useful as they stress the necessity to analyze the data of the problem in the light of the formal, logical structure of it.

*The concept of schema*

Intuitions are very resistant to change. The main reason is that intuitions are related to well structured systems of our cognitive-behavioral, adaptive activity. For that reason, an intuition cannot be changed as an isolated mental device. *Intuitions change together with the entire adaptive system to which they belong*. This is a main hypothesis of our work. In order to understand this claim, one has to resort to the concept of schema.

The adaptive systems mentioned above refer, in fact, to that which is called in the cognitive literature, *schemata*. Let us first try to establish as clearly as possible, the meaning of the term schema – as is referred to above. This is necessary because this term is used with a relatively wide range of interpretations.

Two main interpretations may be distinguished from the beginning. According to an usual, very widespread interpretation, the term schema indicates a kind of condensed, simplified representation of a class of objects or events. The image below is the schema of the human face (see Figure 8). If you have this schema in your mind, you recognize a human face. The sequence of acts in solving a certain class of problems constitute the schema of the solution. For instance, given the problem: The price of

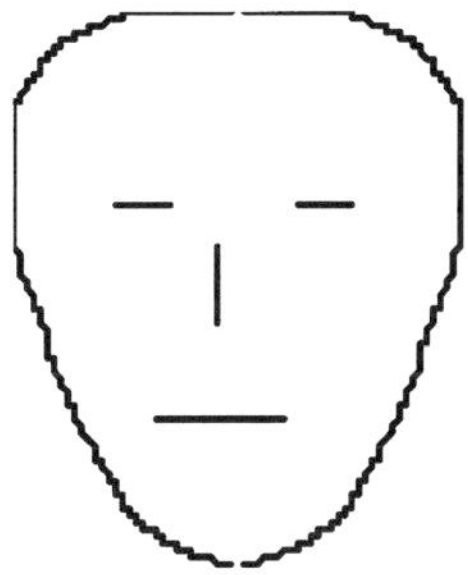

*Figure 8.*

seven pencils is \$ 21. What is the price of ten pencils? One may use the schema of proportion (already mentioned above):

$$\frac{\text{quantity1}}{\text{quantity2}} = \frac{\text{price1}}{\text{price2}}$$

$$\text{price2} = \frac{\text{quantity2} \times \text{price1}}{\text{quantity1}}$$

These are the successive steps for solving the problems, i.e., the schema of the solution.

A second interpretation of the term schema expresses the Piagetian point of view with regard to the adaptive behavior of an organism. As one knows, according to Piaget, the adaptive behavior is achieved through two basic constitutive aspects: assimilation and accommodation. The terms are used with the same meaning in biology and psychology. The food introduced in the organism has to be processed according to certain programs in order to be transformed and integrated in the organism. Such programs are assimilatory schemata. In order to identify a certain object, to understand a text, to solve a problem, one has to *process* the respective information according to adequate assimilatory schemata, and to integrate it in our mental organization. On the other hand, the respective schemata have to accommodate themselves to the specific, particular properties of the respective data. In order to understand a text in English, one has to mobilize the terminology, the grammatical rules of that specific language, to retrieve in our memory the meanings of the respective terms, etc.

The main difference between the first interpretation and the second one is that in the first interpretation, a schema is a limited, specific, executive device, while in the second interpretation a schema plays a general adaptive function in our behavioral-cognitive endeavors. In this second interpretation, a schema represents a precondition, depending on which a person is able to process and integrate a certain amount of information and respond adequately to a class of stimuli. In this interpretation, schemata depend on both the intellectual maturation of the individual and a sufficient amount of training.

Let us consider an example. Would it be possible to teach a 9 year-old child the concept of mathematical proof? Probably not. At this age, the child is concretely oriented; he does not possess the mental tools (conditional reasoning, propositional reasoning, the interpretation of mathematical objects as abstract entities) in order to be able to understand the meaning of a certain mathematical concept. The schema of formal proof may be developed later on together with the global intellectual development of the adolescent. That is: at the age of nine, one cannot teach the child, successfully, a *certain* mathematical proof because the child does not possess, yet, the schema of formal proof, the intellectual device by which the respective concepts and procedures are assimilated, processed, mentally integrated and used efficiently. And, in turn, the schema itself cannot be properly built because the child is not, yet, intellectually, mature in this respect. In principle, the adolescent is potentially prepared to understand the concept of mathematical proof, but this does not happen spontaneously. A certain training is necessary in order to transform the potentiality of the schema into an active, effective assimilatory device.

It will be helpful to quote a number of definitions of the schema concept in order to better suggest the complexity of the notion, and its various interpretations. As a matter of fact, the schema of the schema concept is one of the most complex constructs in cognitive psychology!

Rumelhart mentions that: 'The term *schema* comes into psychology most directly from Bartlett'. Bartlett himself attributes his use of the term to Head (see Rumelhart, 1980, p. 33).

In the 'Dictionary of Psychology' compiled by Arthur Reber (1995) one finds the following: 'A schema is a plan, an outline, a framework, a program'. And, in continuation Reber writes: '... schemata are cognitive mental plans, that are abstract ... they serve as guides for action, as structures for interpreting information, as organized frameworks for solving problems' (Reber, 1995, p. 689).

Flavell, in his important book dedicated to the work of Piaget has written:

> 'A *schema* is a cognitive structure which has reference to a class of similar action sequences, the sequences being strong bounded totalities in which the constituent behavioral elements are tightly interrelated.' (Flavell, 1963, p. 52–53)

And in continuation:

'However elementary the schema, it is schema precisely by the virtue of the fact that the behavior components which it sets into motion, form a strong whole, a recurrent and identifiable figure against less lightly organized behaviors.' (Flavell, ibid. p. 54)

Piaget has written:

'A schema of an action consists in those aspects which are repeatable, transposable or generalisable, i.e., the structure or the form in distinction to the objects which represent its variable content.' (Piaget, 1980, p. 205)

Referring to the general concept of schema, Rumelhart writes that:

'... schemata truly are *the building blocks of cognition*. They are the fundamental elements upon which all information processing depends. Schemata are employed in the process of interpreting sensory data (both linguistic and nonlinguistic) in retrieving information from memory, in organizing actions, in determining goals and subgoals, in guiding the flow of processing in the system.' (Rumelhart, 1980, pp. 33–34)

It is clear from the above quotations how complex the concept of schema is and how large the variety of particular meanings to which it refers.

Let us try to make some order. For this, we will adopt a working definition which tries to synthesize all or most of the characters mentioned above. *A schema is a program which enables the individual to: a) record, process, control and mentally integrate information, and b) to react meaningfully and efficiently to the environmental stimuli.*

From the above definition, one may see that a schema is always *a program* somehow similar to a computer program. The concept of *program* implies that a schema consists in an established sequence of steps leading to a certain purpose. For instance, when I am looking for a certain book at the library, I start by seeking the name of the author (surname, and Christian name) in the catalogue (or with the help of the computer, I start by typing in the name of the author). Secondly, I try to detect the title I am looking for. Thirdly, I have to consider the symbols identifying the book (letters and digits) and to look for the book on the shelves according to these symbols. The *order* of the symbols is also significant, starting from the global domains (psychology, sociology, etc.) to more specific ones, perception, memory, reasoning, motivation and so on. The ensemble of these steps leading, in our example, to a specific book, represents a schema. It is a *learned* schema, not an innate one. Let us also consider examples of inborn schemata: the schema of sucking, of prehension, etc., of the new born child. These are reflexes based on inborn nervous and muscular structures. These, also, are schemata because they represent *sequences of acts according to a certain behavioral program.*

A strategy for solving a certain class of problems constitutes also a *schema*, because it is a *program* enabling the individual to cope efficiently with a certain situation. Such a program of action, such a schema by which one reacts to a certain situation is called by Piaget 'a procedural schema'.

*Specific and structural schemata*

In our opinion, it would be useful to distinguish between specific, content-bound schemata and structural schemata. The specific schemata may also be called *action schemata* because their sequential nature expressed in a string of actions is more evident.

*Examples of specific (action) schemata:* The sequence of steps in arithmetical operations; the sequence of procedures for solving a certain class of problems; the sequence of acts in riding a bicycle (procedural schemata).

The mental processes by which one identifies a real object, or a mathematical object (a geometrical figure, a certain type of number, for instance an irrational number, an equation, a function, constitute 'presentative' schemata in the terminology of Piaget.

*Examples of structural schemata:* The schemata of classification, order, bijection, number, equivalence, empirical and formal proof, proportion, combinatorial capacity, deterministic and probabilistic relationships, causality. One may affirm that structural schema unifies *a principle* with *a program of action*. For instance, the schema of causality implies the general *principle* of causal relationship and the particular identification of the antecedent, the consequent and the necessary relationship between them. The notion of structural schemata proposed by us is a generalization of the Piagetian concept of operational schemata. Operational schemata are in the Piagetian theory, those schemata which are characterized by their *generality*, by their large impact on the individual's cognitive and behavioral activity. Piaget mentions in this respect, the schema of proportion, the combinatorial system, the concept of probability, the concept of inertia, etc. In the Piagetian theory, these schemata emerge during the formal operational stage. All these constitute in our terminology specific schemata.

*Every stage* in the intellectual development of the child is characterized by a number of schemata with very large implications for his behavior and intellectual capacity. We have called them *structural schemata*. For instance, the concrete operational stage is characterized by the conservation capacity, by the assimilation and use of the concept of number and the operations with numbers, by the concept of mechanical causality, etc.

### *Intuitions and schemata: The evolution of intuitions with age*

Intuitions and schemata are two different types of cognitive categories. Intuitions are cognitions characterized mainly by globality and the subjective feeling of obviousness. Schemata, as mentioned above, are programs of interpretation and reaction. What these two categories have in common is their basic role in the adaptive processes, their profound connections with the structural capacities of the individual. But cognitive psychology has not, so far, considered the possible connection of these two categories in the realm of the adaptive behavior. *Recently, a number of experimental findings lead us to the assumption that intuitions are profoundly related to structural schemata (as defined above) and that both categories display a certain evolution as an effect of age*. That evolution may be explained by a particular type of interaction between schemata and intuitions. Let us first mention the facts.

### *The conservation capability and its evolution with age*

One may distinguish between pre-operational and operational intuitions (evidently related to the respective evolution periods). Let us remember one example:

A 5 year-old child is presented with two identical containers (A and B). The child is asked to introduce pairs of marbles, one in each container, successively. Asked to compare the quantity of marbles in the two containers, the child affirms that the amount is the same. His answer has the characteristics of an intuitive cognition. It is direct, global, self-evident.

One pours the marbles from one of the containers, let's say, B, into container $B^1$, having a different shape than B: taller and narrower. The child, again asked to compare the amount of marbles, will answer, usually, that in $B^1$ there are more marbles than in A. The 5 year-old child does not 'conserve' (in the Piagetian terminology). His estimation is still intuitive: direct, global, self-evident.

The same questions are put to a 6–7 year-old child. The two amounts of marbles in A and in $B^1$ are considered to be identical. Again the answer is direct, self-evident. The child is surprised that he has been asked such a question (Piaget, Inhelder and Szeminska, 1960). The intuition has changed. The non-conservation intuition is a pre-operational intuition. The conservation intuition is an operational intuition. The non-conservation intuitions, in general, are related to the ways in which the pre-operational child interprets reality, through his intellectual schemata. In the above example, the interpretation is based on a global configuration determined by a single dimension (height of the container). The interpretation expresses a certain lack of reversibility. In the older child, *two* dimensions intervene (height and width). The invariance (based on reversibility) is taken into account. These are intellectual schemata, programs through which the original information is acquired, processed and integrated. The final (the intuitive) interpretation is sudden, global, apparently obvious, but in fact, based on sequences of transformations and integrative steps.

Based on the above examples, one may assume that intuitions are related to intellectual schemata. As intellectual schemata change, as an effect of age and experience, corresponding intuitions change too. Certainly, this is only an assumption. It has to be checked in respect of various circumstances. From the above examples, one may already assume the existence of pre-operational intuitions – related mainly to the period of intuitive thinking (in the Piagetian terminology) – and of operational intuitions, starting at the age of six – seven – (the concrete and the formal operational thinking). During the concrete operational period, one may identify intuitive cognitions specifically emerging during this period, for instance, the intuition of number, the intuitions related to Euclidean geometry, the social understanding of time measurements, the intuitive understanding of mechanical causality, the intuitive distinction between deterministic and stochastic relationships.

An experiment performed by Wilkening testified the role of implicit schemata in intuitive evaluations. Wilkening asked children and adults to evaluate, by comparison, areas of rectangles. Using a statistical procedure based on analysis of variance, Wilkening found that 5 year-old children used tacitly an additive rule (addition of the two dimensions of the rectangle), adults used the multiplicative rule, while 8 and 11 year-old children used either the additive or the multiplicative rule (Wilkening, 1980, pp. 54–58).

These computational schemata express, in fact, more general, more profound – that is, structural – schemata: an additive (more primitive) or a multiplicative (a higher order) schemata. These are structural ways of reasoning. We emphasize that these computational schemata act tacitly: The corresponding, explicit intuition is expressed in a cognitive, intuitive, estimation.

A more recent research performed by Clark and Kamii (1996) studied the effect of age on the two basic arithmetical structures: addition and multiplication. Usually, one defines multiplication (of natural numbers) as repeated addition. But Clark and Kamii show that the two arithmetical operations reveal two different intellectual schemata, with the idea of addition emerging first and the concept of multiplication developing later on. Intuitive responses reflect this evolution.

Let us now consider two cognitive – mathematical domains in which the relationship of schemata and intuitions have been explored: infinity and probability.

*Infinity:*
In a previous research (Fischbein, Tirosh and Hess, 1979), we addressed the following question: 'How does age affect the understanding of mathematical infinity?' The subjects were students enrolled in grades 5 to 9. Questions appearing in a written questionnaire, asked the subjects to compare a) the amount of points in two line segments of different lengths, and b) in a line segment and in a square or a cube. Other questions asked the subjects to indicate whether the successive divisions of a line segment by two and by three will come to an end or not, to compare the set of natural numbers with the set of positive even numbers, etc.

It has been found that the percentages of 'finitist' and 'infinitist' answers, oscillate across age (grades) for some questions. Despite this, the general trend remains relatively stable across age. For instance, with regard to successive divisions by three, 38% of the subjects in grade 5 and 36.7% of subjects in grade 9 claimed that the process is infinite. On the other hand, about 60% in grade 5 and 62.4% in grade 9 claimed that the process come to an end.

One of the items raised the following question: 'Consider a square and a cube. Is it possible to find a point of correspondence on the square for each point of the cube?' The notion of one-to-one correspondence had been previously explained.

Let us consider first the effect of age. Surprisingly, *the percentage of the correct answers (the equivalence of the two sets of points) decreases with age*. The percentages of the mathematically correct answers (the equivalence of the two sets of points) are: 30.3% in grade 7; 27.9% in grade 8 and 21.3% in grade 9. Correspondingly, the percentages of the mathematically incorrect answers increase with age.

Let us consider another item of the same type: The subjects were asked whether one-to-one correspondence may be established between the points of a line segment and the points of a square. The data presented the same picture as above. *The proportion of correct answers (equivalence of the two sets of points) decrease with age*, namely: 32.9% in grade 7, 31.7% in grade 8 and 17.6% in grade 9. Correspondingly, the proportion of incorrect answers (absence of equivalence) has been found to increase with age (grade).

How can these findings be explained? In our opinion, the intuitions discussed express certain intellectual schemata. In the present case, one has to consider the principle: 'The whole is bigger than a part of it' which is consistent with our intuitive estimations. An intuition is never a mere guess. As we assumed above, an intuition depends on a structural schema. *As an effect of age, the intervention of this logical principle is applied more and more consistently as one may naturally expect.* The problem is that the respective intellectual schema is adequate for finite entities, but not for infinite entities. The intuitive answer is manipulated from 'behind the scenes' by this non-adequate principle. *As an effect of age, more subjects will yield the mathematically incorrect answer.*

Generally speaking, one may conclude the following:
Intuitions (global, self-evident views) are the cognitive counterpart of some structural, intellectual, schemata. Structural schemata develop with age, becoming more consistent or more efficient. Intuitions may, sometimes, be related to adequate schemata, but, sometimes, they may be manipulated by non-adequate schemata. In this second type of situation, as an effect of the natural improvement with age of the respective schemata (consistency, efficiency), *the corresponding intuitive evaluations will worsen as an effect of age.*

*Probabilistic intuitions:*
A conformation of the hypothesis that intuitions are related to intellectual schemata has been obtained in a recent research of ours referring to the evolution with age of probabilistic intuitions (see Fischbein and Schnarch, 1997).

It has been found that intuitions develop with age in a divergent manner. Some intuitions improve with age (decrease of the proportion of misconceptions). Some intuitions become worse with age (increase in the proportion of misconceptions), while certain intuitions remain stable.

Let us start with a finding already mentioned above (see page 25):

Subjects were asked to compare the probability of two groups of numbers in a lotto game (I) 1, 2, 3, 4, 5, 6 versus (II) 39, 1, 17, 33, 8, 37. Comparing the probabilities of group I versus group II, one may get:

a) $P(I) > P(II)$;
b) $P(I) < P(II)$;
c) $P(I) = P(II)$

None of the students chose answer (a). Answer b) (main misconception) was chosen by 70% (grade 5); 55% (grade 7); 35% (grade 9), and 22% (grade 11). The adequate principle in this example is that of independence of probabilities, about which subjects have become more and more aware. *Correspondingly, the proportion of correct answers – equality of chances – increased with age.*

Let us now consider an opposite example:
The likelihood of getting heads at least twice when tossing three coins is:

a) Smaller than ... (incorrect)
b) Equal to ... (incorrect – main misconception)
c) Greater than ... (correct) the likelihood of getting heads at least 200 times out of 300 times.

It has been found that the response (b) (equality of chances), *which is erroneous*, has been given by the following proportions of subjects: 30% (grade 5); 45% (grade 7); 60% (grade 9) and 80% (grade 11).

In this case, the subjects based their intuitive evaluation on a non-adequate schema: the schema of proportion. They did not consider the magnitude of the samples. The principle which should have been applied is the following: As an effect of increasing the samples, the empirical probability approaches the theoretically predicted probability. This is called 'the law of large numbers'. Consequently, the likelihood of getting at least 200 heads when tossing a coin 300 times is very small, certainly smaller than the likelihood of getting 2 heads, when tossing a coin three times. But the subject's attention is in this case captured by the most salient data, that is, the schema of proportion. Justifying their answer, most of the subjects affirmed that: the chances are the same, because $\frac{200}{300} = \frac{2}{3}$.

The schema of proportion is, in the Piagetian terminology, an operational schema, that is, a very general, a very influential one. It develops with age and in its full, quantitative form, it manifests itself during the formal operational stage. Together with other intellectual schemata, the schema of proportion deepens its impact on the individual's reasoning, as an effect of age.

*But in the above problem, the idea of proportion is misleading.* The schema is not adequate, and consequently, the percentage of erroneous answers – the equality of probabilities – increases across ages.

Let us consider the following problem (see Fischbein and Schnarch, 1997). Yoav and Galit receive each a box containing two white marbles and two black marbles.

A. Yoav extracts a marble from his box and finds out that it is a white one. Without replacing the first extracted marble, he extracts a second marble. The likelihood that this second marble is also white is smaller, equal or bigger compared to the likelihood that it is a black marble?

Explain your answer.

B. Galit extracts a first marble from her box and puts it aside without looking at it. She then extracts a second marble and sees that it is

TABLE I

Problems and percentages of student answers

| Problems | Grades | | | |
|---|---|---|---|---|
| | 5 | 7 | 9 | 11 |
| Category 1 (both correct) | 45 | 50 | 35 | 30 |
| Category 2 (the first correct, the second incorrect) (the main misconception) | **5** | **30** | **35** | **70** |
| Category 3 (both incorrect: equality of chances for) both questions) | 25 | 15 | 25 | 0 |
| Others | 25 | 5 | 5 | 0 |

Main misconception responses are highlighted.

> white. Is the likelihood that the first extracted marble is white, smaller than, equal to, or bigger than the likelihood that it is black?

The results are shown in Table I.

The table divides the answers into three categories. In the first category, both answers are correct. For the first question, obviously, the correct answer is that after extracting a white marble without replacement, the chances of extracting another white marble are smaller (two black and one white remained).

With regard to the second question, the problem is much more subtle: Galit extracts a first marble and without looking at it, puts it aside. She extracts a second marble *and sees that it is white*. Is the likelihood that the first extracted marble is white, smaller, equal or greater than the likelihood of it being black? One deals with the schema of the time axis (or the cause-effect axis). *As an effect of age* more and more subjects claim that at the first extraction *the chances were equal*, no matter what happened with the second extraction.

Again, the subjects who answered this way are influenced by a non-adequate schema. The problem is not a physical one, a problem of successive events. The problem is merely a cognitive one. *One knows* that with the second extraction, one has drawn a white marble. *This information* determines the answer.

What the subjects do not seem to realize is that the *knowledge* of the second outcome should be used in determining the probability of the first outcome. If we know that the second outcome is a white marble, then the remaining three from which the first one has been drawn are two black and one white. The erroneous intuition is caused by the tacit embedding of the principle of causality (physical antecedent $\rightarrow$ consequent relation) in the intuitive evaluation. Imagine that the above experiment (phase B) had been repeated many times and all the outcomes have been discarded in which a white marble has been drawn with the second extraction, then only the set of two black and one white should be considered in the final stochastic evaluation.

Briefly speaking, in this example as well, one finds that the emerging intuition is controlled, in fact, tacitly by a principle, an intellectual attitude, a structural schema. And, in general, if the intervening schema is adequate to the data of the problem, one should find that the respective intuitions improve with age, because structural schemata progress with age. But, if, on the contrary, the schema – elicited by some salient (but not essential) data and the respective schema is not adequate to the essence of the problem, one should find that the intuition worsens with age: *The progress of the schema only deepens its non-adequate influence upon the intuitive solution leading to an increase, across ages, of the proportion of erroneous intuitions*.

### *The process of compression*

In two recent studies, we addressed the following question: What is the relationship between the intuitive estimation of the number of possibilities in a combinatorial problem, on one hand, and the correct mathematical solutions? Groups of 3, 4, and 5 objects were considered.

The following types of combinatorial problems were studied: Permutations, arrangements with and without replacement and combinations. It has been found that the subjects tend to underestimate the number of permutations (except the case of 3 objects) and overestimate the number of arrangements with and without replacement and in problems with combinations. Two findings were especially interesting:

- The order of magnitudes of the intuitive evaluations followed the order of magnitudes of the correct mathematical solutions – for each of the groups of objects (3, 4, and 5 objects). Mathematically, the order of magnitudes is the following: Arrangements with replacement > Arrangements without replacement > combinations separately for each of the groups of objects (3, 4, and 5). *The same order has been found when considering the intuitive estimations* (see Fischbein and Grossman, 1997; Zamir, 1996). The intuitive estimations, though apparently spontaneous, are nevertheless controlled by objective considerations.
- The second finding was the following: The subjects were initially invited to produce a global, spontaneous estimation of the number of possibilities for each combinatorial problem – that is, without any

computation. Some of the subjects were subsequently interviewed and asked to explain how they reached their answer. This certainly poses a methodological problem. It is possible that some of the subjects remember an operation they really performed when trying to process their answer. But it is also possible that some subjects simply imagined a posteriori an operation which should fit their guess. It is also possible that the subjects themselves were not aware of what really happened. *All the subjects reported a binary operation, a multiplication, no matter the number of operations which should have been performed according to the respective formula.* The fact that *all* the subjects reported a multiplication when dealing with a combinatorial problem, supports our claim that an intuitive guess is based on a structural schema. The combinatorial system is essentially a multiplicative one. Secondly, the operations reported were based on the numbers characterizing the problem. For instance, referring to an arrangement with replacement of 4 elements taken by 2, the subjects reported $2^4$ or $4^2$ or $2 \times 4$. The same operations were reported for arrangements of 4 elements taken by 2 without replacement, or for combinations of 4 elements taken by two. That is, no matter the number of operations really implied by the correct combinatorial computation, the subjects a) considered the numbers which were directly salient in the text of the problem, and b) always reduced the computation to a binary operation. In other terms: in their combinatorial intuitive guess, the subjects a) took into account some of the most directly visible data; b) put into action (correctly or incorrectly) a certain structural schema (in the present case, the multiplicative schema) together with some specific computational schema, and finally, c) *compressed* the entire sequence of operations to a minimal operation which, in the present case, was a binary multiplication.

The phenomenon of compressions seems to have a fundamental role in the mechanisms of intuition. *One may assume that, generally, the transition from a schema, which is a sequential process, to an intuition that is a global, apparently sudden, cognition is achieved by a compression process.*

Compression, as Thurston (1990) has shown, has a very general function in mathematical reasoning. ‘Mathematics is amazingly compressible: you may struggle a long time, step by step to work through some process or idea from several approaches. But once you really understand it and have the mental perspective to see it as a whole, there is often a tremendous mental compression’ (Thurston, 1990, p. 887). One may assume that compression does not lead, necessarily, to an intuition: symbols, formulas, theorems represent compressed mathematical entities, but, usually, they do not have an intuitive meaning. On the other hand, an intuition is, generally, the effect of a compression, if a structural schema lies behind this cognition.

As we have seen above, the structural schema may not be adequate. The compression process to which the schema is subjected, will lead, in this case, to an intuition, but not to a correct one.

## REFERENCES

Balzan, M.: 1997, 'Difficulties in Understanding the Concept of Sets', Masters Thesis, Tel Aviv University, Tel Aviv, Israel (in Hebrew).

Bastick, T.: 1982, *Intuition: How we Think and Act*, John Wiley & Sons, Chichester, England.

Bergson, H.: 1954, *Creative Evolution* (translated by A. Mitchell), Macmillan, London.

Bruner, J.: 1965, *The Process of Education*, Harvard University Press, Cambridge, MA.

Bunge, M.: 1962, *Intuition and Science*, Prentice Hall Inc., Englewood Cliffs.

Carey, S.: 1985, *Conceptual Change in Childhood*, MIT Press, Cambridge, MA.

Clark, F. B. and Kamii, C.: 1996, 'Identification of multiplicative thinking in children in grades 1–5', *Journal for Research in Mathematics Education* 27, 45–51.

Clement, J., Brown, D. E. and Zietsman, A.: 1989, 'Not all preconceptions are misconceptions: Finding 'anchoring conceptions' for grounding instruction on students' intuitions', *International Journal for Science Education* 11, 554–565.

Descartes, R.: 1967, *The Philosophical Works* (Vol. 1), (translated by E. S. Haldane and G. R. T. Ross), The University Press, Cambridge, MA.

DiSessa, A. A.: 1988, Knowledge in pieces, in G. Forman and P. B. Putall (eds.), *Constructivism in Computer Age*, Erlbaum, Hillsdale, NJ, pp. 49–70.

Fischbein, E.: 1963, *Conceptele Figurale* [The figural concepts], in Roumanian Bucharest, Editura Academiei, R.S.R.

Fischbein, E.: 1975, *The Intuitive Sources of Probabilistic Thinking in Children*, D. Reidel, Dordrecht, The Netherlands.

Fischbein, E.: 1978, 'Schèmes virtuels et schème actifs dans l'apprentissage des scinces', *Revue Française de Pédagogie*, 119–125.

Fischbein, E.: 1987, *Intuition in Science and Mathematics*, D. Reidel, Dordrecht, The Netherlands.

Fischbein, E. and Grossman, A.: 1997, 'Schemata and intuitions in combinatorial reasoning', *Educational Studies in Mathematics* 34, 27–47.

Fischbein, E. and Nachlieli, T.: 1997, *Concepts and Figures in Geometrical Reasoning*, submitted.

Fischbein, E. and Schnarch, D.: 1997, 'The evolution with age of probabilistic, intuitively based misconceptions', *Journal for Research in Mathematics Education* 28, 96–105.

Fischbein, E., Tirosh, D. and Hess, P.: 1979, 'The intuition of infinity', *Educational Studies in Mathematics* 10, 3–40.

Fischbein, E., Tirosh, D. and Melamed, U.: 1981, 'Is it possible to measure the intuitive acceptance of a mathematical statement?' *Education Studies in Mathematics* 12, 491–512.

Flavell, J.: 1963, *The Developmental Psychology of Jean Piaget*, Van Nostrand, Reinhold Co., New York.

Gelman, R. S. and Gallistel, C. R.: 1978, *The Child's Understanding of Number*, Harvard University Press, Cambridge, MA.

Hahn, H.: 1956, 'The crisis of intuition', in J. Newman (ed.), *The World of Mathematics*, Simon and Schuster, New York, v.3, pp. 1957–1976.

Kant, I.: 1980, *Critique of Pure Reason*, (translated by N.K. Smith), Macmillan, London.

Kline, M.: 1980, *Mathematics: The Loss of Certainty*, Oxford University Press, New York.

McCloskey, M., Washburn, A. and Felch, L.: 1983, 'Intuitive physics. The straight-down belief and its origin', *Journal of Experimental Psychology: Learning, Memory and Cognition* 9, 636–649.

Piaget, J.: 1980, *Experiments in Contradiction*, University of Chicago Press, Chicago and London.

Piaget, J., Inhelder, B. and Szeminska, A.: 1960, *La Géometrie Spontanée de l'Enfant*, Press Universitaire de France, Paris.

Reber, A. S.: 1995, *Dictionary of Psychology*, Penguin Books, London.

Resnick, L. B.: 1987, 'The development of mathematical intuitions', in M. Perlmater (ed.), *Minnesota Symposium of Child Psychology*, Erlbaum, Hillsdale, NJ, p. 159–194.

Rumelhart, D. E.: 1980, 'Schemata: The building blocks of cognition', in R. T. Spiro, B. C. Bruce and W. F. Brewer (eds.), *Theoretical Issues in Reading Comprehension*, Erlbaum, Hillsdale, NJ.

Spinoza, B.: 1967, *Ethics and Treatise on the Correction of the Understanding*, (translated by A. Boyle), Everyman's Library, London.

Stavy, R. and Tirosh, D.: 1996, 'The role of intuitive rules in science and mathematics education', *European Journal of Teacher Education*, 19, 109–119.

Stavy, R., Tirosh, D. and Tsamir, P.: 1997, 'Intuitive rules and comparison tasks: The grasp of vertical angles', in G. Makrides (ed.), *Mathematics Education and Applications*, Cyprus Pedagogical Institute, Nicosia, Cyprus, pp. 269–276.

Thurston, V. P.: 1990, 'Mathematical education', *Notices of the American Mathematical Society* 137, 850–884.

Tirosh, D.: 1991, The role of students' intuitions of infinity in teaching the Cantorian theory, in D. Tall (ed.), *Advanced Mathematical Thinking*, Kluwer, Dordrecht, The Netherlands, pp. 199–214.

Westcott, M. R.: 1968, *Towards a Contemporary Psychology of Intuition*, Holt Rinehart and Winston, New York.

Wilkening, F.: 1980, 'Development of dimensional integration in children's perceptual judgment: Experiments with area, volume and velocity', in F. Wilkening, J. Becker and T. Trabasco (eds.), *Information Integration by Children*, Lawrence Erlbaum, Hillsdale, NJ.

Zamir, T.: 1996, 'Combinatorial intuitions', Masters Thesis, Tel Aviv University, Tel Aviv, Israel (in Hebrew).

PROFESSOR EFRAIM FISCHBEIN†
*School of Education,*
*Tel Aviv University,*
*Tel Aviv 69978,*
*Israel*

DINA TIROSH and RUTH STAVY

# INTUITIVE RULES: A WAY TO EXPLAIN AND PREDICT STUDENTS' REASONING

ABSTRACT. Through our work in mathematics and science education we have observed that students react similarly to a wide variety of conceptually unrelated situations. Our work suggests that many responses which the literature describes as alternative conceptions could be interpreted as evolving from common, intuitive rules. This paper describes and discusses one such rule, manifested when two systems are equal with respect to a certain quantity A but differ in another quantity B. We found that in such situations, students often argue that 'Same amount of A implies same amount of B'. Our claim is that such responses are specific instances of the intuitive rule 'Same A–same B'. This approach explains common sources for students' conceptions and has strong predictive power.

## 1. Intuitive rules: A way to explain and predict students' reasoning

In the last decades, researchers have studied students' conceptions and reasoning in the context of mathematics and science education. Many have pointed out the persistence of alternative conceptions which are not in line with accepted scientific notions. Such conceptions cover a wide range of subject areas. Most of this research has been content-specific and aimed for detailed descriptions of particular alternative concepts. Yet, there is evidence that students tend to respond inconsistently to tasks related to the very same mathematical or scientific concepts (Clough and Driver, 1986; Nunes, Schliemann and Carraher, 1993; Tirosh, 1990). This constitutes a challenge to the alternative conception paradigm.

Through our work in mathematics and science education, we have observed that students react in a similar way to a wide variety of conceptually nonrelated problems which share some external, common features. These problems differ with regard to their content area and/or to their required reasoning. For instance, students' responses to comparison tasks embedded in different content areas are often of the type: 'More of A–more of B' (Stavy and Tirosh, 1996). An example is offered by studies on the development of the concept of temperature. These studies showed that when children were presented with two cups of warm water, one containing

*Educational Studies in Mathematics* **38:** 51–66, 1999.

twice as much water as the other, they claimed that 'the more water–the warmer' (Erickson, 1979; Stavy and Berkovitz, 1980). This response is often interpreted as an alternative conception of temperature. Another example relates to children's conceptions of angle. Noss (1987) presented children with two identical angles, one of which had 'longer arms' than the other. They found that many children between the ages of ten and fifteen argued that 'the angle with the longer arm is bigger'. This response was interpreted as an alternative conception of angle.

Within the framework of intuitive rules, we interpret such responses as evolving from a common source, namely the intuitive rule 'More of A–more of B'. This intuitive rule is reflected in students' responses to many comparison tasks, including classical, Piagetian conservation tasks (conservation of number, area, weight, volume, matter, etc.) and tasks related to intensive quantities (density, temperature, concentration, etc.). All these tasks share some common features. In each of them, two objects (or two systems) which differ in a certain, salient quantity A are described ($A_1 > A_2$). The student is then asked to compare the two objects (or systems) with respect to another quantity B ($B_1 = B_2$ or $B_1 < B_2$). In all these cases, a substantial number of students responded inadequately according to the rule 'More of A (the salient quantity)–more of B (the quantity in question)' arguing that $B_1 > B_2$. In general, we argue that many alternative conceptions, apparently related to specific domains, are actually only specific applications of this rule (Stavy and Tirosh, 1996).

The intuitive rule 'More of A–more of B' is activated by salient, perceptual differences between two objects (or systems) in respect to a certain quantity A. However, we have recently noticed that when $A_1 = A_2$ and $B_1 \neq B_2$, students often claim that $B_1 = B_2$. Our claim is that such responses are specific instances of another intuitive rule: 'Same amount of A–same amount of B (Same A–same B, for short).' This paper deals with this intuitive rule. We shall first refer to situations in which the equality in quantity A is directly given in the task. Then, to situations in which the equality in quantity A is not directly given but can be deduced logically.

## 2. Directly-Given Equality

We shall start this section with two cases which have been previously presented as alternative conceptions in the mathematics or science education literature. We interpret these responses as specific instances of the use of the rule 'Same A–same B'.

*Length and distance.* Piaget, Inhelder and Szeminska (1960) asked young children to compare the length of a straight line with that of a wavy line. The lines were of different lengths but they began and finished at parallel points on the page. Piaget et al. (1960) reported that 84% of children aged four to five incorrectly replied that the lines were equal in length. Piaget et al interpreted this response by referring to children's development of the concept of length. They argued that 'at this stage, the length of a line is estimated solely in terms of its endpoints without reference to its rectilinearly' (pp. 92). Clearly, however, we may also regard this response as a case in which the intuitive rule 'Same A (distance between endpoints)–same B (length of lines)' is activated.

*Concentration and temperature.* Children aged four to fourteen were presented with two cups of water and were asked about the relative sweetness of the water after sugar was added to the cups. One cup was full of water and one teaspoon of sugar was mixed into it. The same was done with the other, same sized, but half–full, cup. The children were asked whether they thought the sweetness of the sugar water in the two cups was the same or not, and if not, in which cup the water was sweeter.

This task was included in a study on the development of children's conception of concentration (sweetness) conducted by Stavy, Strauss, Orpaz and Carmi (1982). Most of the young participants (four to eight year olds) argued that each cup contains one teaspoon of sugar and water and, as a consequence, they must be equally sweet. Very similar results were obtained in regard to the development of children's conceptions of temperature (Stavy and Berkovitz, 1980; Strauss and Stavy, 1982).

The behaviors of the young children in these studies were often interpreted as the application of incorrect alternative conceptions related to nondifferentiation between mass and concentration or nondifferentiation between heat and temperature (Erickson, 1979; Wiser and Carey, 1983). Another explanation related to children's difficulty in coping with inverse ratio in the context of intensive quantities (Strauss and Stavy, 1982; Noelting, 1980a, 1980b). Our claim is that such incorrect responses could also be viewed as applications of the general rule 'Same A (same amount of sugar)–same B (same sweetness)'.

So far, we have shown that the rule 'Same A–same B' can explain students' responses that were previously described as alternative conceptions in the mathematics and science education research literature. In order to test the predictive power of this rule, we specifically designed several tasks. In these tasks, students were presented with two objects (or systems), which were equal in respect to a certain quantity A ($A_1 = A_2$), but differed in another quantity B ($B_1 > B_2$ or $B_1 < B_2$). Students were asked to compare $B_1$ and $B_2$.

*Rate of cooling.* Livne (1996) presented students with the following task:

*Debby is a baby-sitter. The baby she is watching wakes up crying, and Debby wants to feed her. She realizes that the milk she has heated is too hot, and wants to cool it as fast as possible. She has two differently-shaped bottles, a ball-shaped one and a cylinder-shaped one, Each bottle can contain 100 ml. She fills each bottle with 100 ml of milk (up to the nipple), and immerses them in ice water. What is your opinion? Is the time needed to cool the milk in the ball-shaped bottle equal/ not equal to the time needed to cool the milk in the cylinder-shaped bottle? If you think that the time is not equal, in which bottle does the milk cool faster? Why?*

Clearly, the cylindrical bottle will cool faster as the rate of cooling depends on the ratio surface-area to volume. In this task the amount of milk in the two bottles is the same ($A_1 = A_2 = 100$ ml) and therefore we predict that a substantial number of students will incorrectly judge that 'same amount of milk–same rate of cooling'.

This problem was presented to biology-major students in grades 10, 11 and 12. Less than 50% of the students in each grade level knew that the cylindrical bottle would cool faster, basing their judgment on the ratio between surface area and volume. However, as predicted, a nonnegligible number of students (29%, 31% and 38% of the students in grades 10, 11 and 12 respectively) argued that 'The time needed to cool the milk in both bottles is equal *because* the amounts of milk in each bottle are equal'. Although these students have learnt about the ratio of surface area to volume and its role in biological systems, their response was of the type 'Same A (amount of milk)–same B (rate of cooling).' This behavior of mature students with a relatively high level of biology education suggests that the rule 'Same A–same B' has coercive power.

In this case, the equality in quantity A was expressed by a number (100 ml). In the next task, the equality in quantity A is mentioned, but no numbers are given in the problem:

*Angles in polygons.* Roghani (1997) presented students in grades 4 to 12 with the following problem:

*Consider a pentagon and a hexagon. All the sides of the pentagon are equal. All the sides of the hexagon are equal. The side of the pentagon is equal to the side of the hexagon.*

TABLE I

Distribution of responses to the comparison of angles in a hexagon and a pentagon (in %)

| Judgment | Grades | | | | |
|---|---|---|---|---|---|
| | 4 (n = 74) | 6 (n = 67) | 8 (n – 64) | 10 (n = 59) | 12 (n = 57) |
| a* | 19 | 30 | 27 | 25 | 46 |
| b | 10 | 5 | 13 | 25 | 30 |
| c (same–same) | 55 | 57 | 50 | 32 | 16 |
| d | 5 | 6 | 8 | 18 | 6 |
| No answer | 11 | 2 | 2 | 0 | 2 |

(*) correct answer

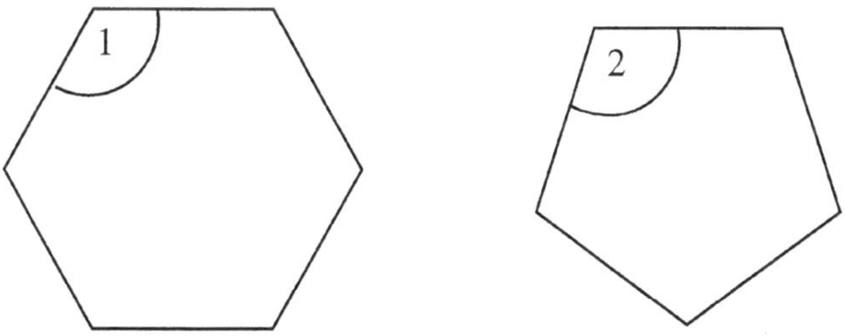


*Circle your answer:*

*a. Angle 1 is greater than Angle 2.*
*b. Angle 2 is greater than angle 1.*
*c. Angle 1 is equal to angle 2.*
*d. It is impossible to determine.*

*Explain your choice.*

Angle 1 is greater than Angle 2. However, we predicted that due to the equality in the sides of the two polygons and/or to the overall similarity between the two drawn objects, many students would claim that 'same sides/object–same angles'.

Table I shows that, as predicted, about 50% of the students in grades 4, 6 and 8 and a substantial number of 10th and 12th graders incorrectly claimed that the two angles were equal. The most typical justification of the younger students (grades 4–8) was: 'The sides are equal, so the angles are equal'. This response directly reflects application of the intuitive rule 'Same A–same B' to this specific situation. The older students, who had studied Euclidean geometry, used more elaborate justifications. e.g., 'In a triangle, the angles opposite equal sides are equal', or 'If each of the two polygons were to be bounded by a circle, the chords in these two circles would be equal, and, in accordance with the theorem related to angles and equal chords, the angles would be equal as well'. These students used geometrical, theorem-like statements, all in line with the intuitive rule 'Same A–same B'. The evident result is overgeneralization of geometrical theorems.

In the following example no quantitative equality was involved, but the two objects shared a qualitative property (i.e., shape).

*Surface Area/Volume of Cubes.*

*Consider two differently-sized cubes. Is the ratio between the surface area and volume of Cube 1 larger than/equal to/smaller than the ratio between the surface area and volume of Cube 2? Explain your answer.*

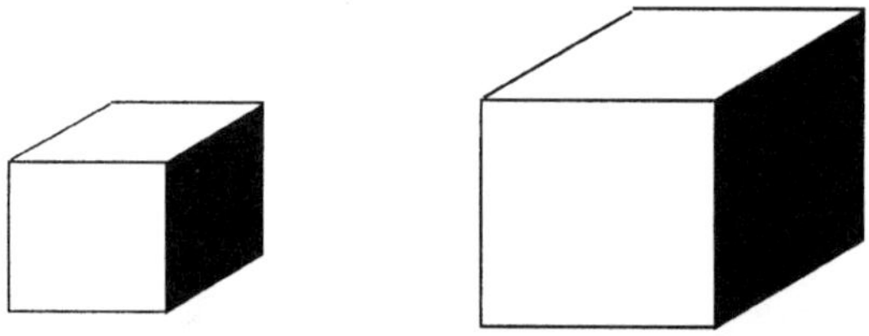

1 2

In this task, the ratio surface area/volume of Cube 1 is larger than the ratio surface area/volume of Cube 2. Again, our prediction was that students would be affected by the identity in shapes and would claim that 'same shape (cube), same ratio (surface area/volume)'. This task was included in Livne's (1996) study, whose participants were biology majors in grades 10, 11 and 12. As predicted, substantial percentages (41%, 45% and 55% in grades 10, 11 and 12 respectively) incorrectly argued that the ratio surface area/volume in the cubes is the same. Typical explanations were: 'Cube 1 and cube 2 have the *same geometrical shape*, hence *the ratio* of surface area to volume in both cubes *is the same* regardless of their size', 'The surface area and the volume in Cube 1 are proportionally smaller than in Cube 2 and therefore the ratio is constant'. In the first justification, students explicitly referred to the shared qualitative property–shape. In the second justification, 'formal' schemes such as proportion were integrated in an attempt to support the judgment.

These last three examples demonstrate the strong predictive power of the intuitive rule 'Same A–same B'.

In all tasks described so far, the equality in quantity A was explicitly given (perceptually, in numerical or verbal terms, or in qualitative properties). Based on these examples, one may assume that the intuitive rule

'Same A–same B' is activated in comparison tasks by the explicit presentation of the equality in quantity A. We predict that the intuitive rule 'Same A–same B' will also be activated when the equality in quantity A is not directly given but is logically deduced. The next section examines this hypothesis.

## 3. Logically-Deduced Equality

Two schemes which can lead, in certain tasks, to equality judgments, are conservation and proportion. It is widely documented in the literature that these schemes develop with age. We shall first relate to the conservation scheme.

### 3.1. *Conservation*

*Surface area and volume of two cylinders*
Consider the following task:

*Take two identical rectangular (non-square) sheets of papers (Sheet 1 and Sheet 2):*

- *Rotate one sheet (sheet 2) by 90°*

  *a. Is the area of Sheet 1 equal to/larger than/smaller than/ the area of Sheet 2?*

- *Fold each sheet (as shown in the drawing). You get two cylinders: Cylinder 1 and Cylinder 2.*

  *b. Is the volume of Cylinder 1 equal to/larger than/smaller than/ the volume of Cylinder 2?*

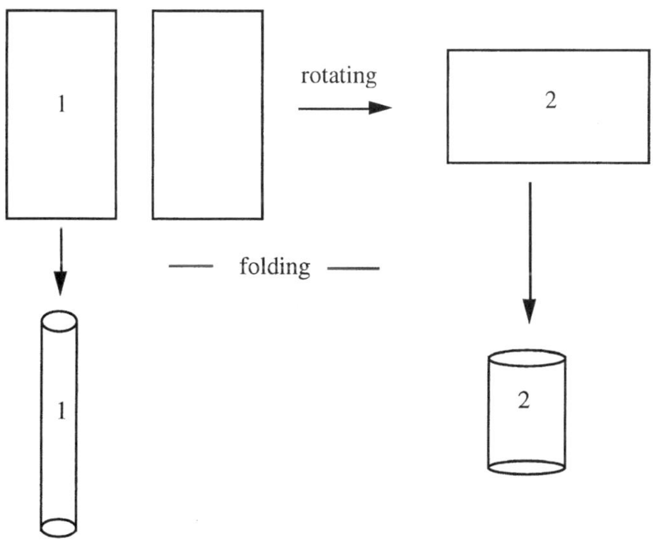

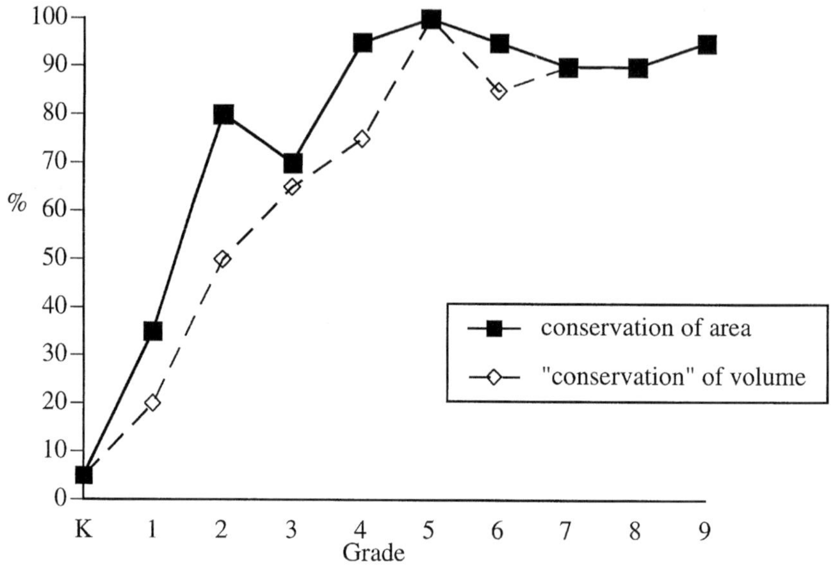


*Figure 1.* Distribution of equality judgment, by age, to the task of surface area and volume of two cylinders

The first question is a classical, Piagetian, conservation of area task. Clearly, the areas of the two sheets are equal ($S_1 = S_2$). The second question deals with another variable involved in this manipulation, namely the volume of the cylinders. The volume of Cylinder 2 is larger than that of Cylinder 1.

We expect, in line with the rule 'Same A–same B', that when students start to conserve the area they argue that the volumes of the two cylinders are equal as well.

To test this hypothesis, we presented K to 9th grade students with the task described above (Ronen, 1996). Students' responses are presented in Figure 1.

Our findings in respect to the conservation of area task are similar to those reported by Piaget, Inhelder and Szeminska (1960). As can be seen from the conservation developmental curve in Figure 1, high percentages of kindergartners and first graders (95% and 65% respectively) incorrectly argued that the areas of the rectangles are not equal, arguing that 'the longer (the side)–the larger (the area)' or 'the taller (the side)–the larger (the area)'. From second grade on, the vast majority of the children conserved the area. Most of them based their judgments on identity, reversibility, additivity and compensation arguments (e.g. 'It's the same paper'; 'You can turn it back and see that it's the same', etc.). This change in behavior is explained by Piaget et al. (1960) as resulting from the development of the logical conservation scheme.

Figure 1 also shows a gradual increase from K to Grade 5 in the percentages of students who incorrectly argue that the volumes of the two cylinders are equal ('conservation' of volume). From Grade 5 on, almost all students (more than 85%) incorrectly claimed that the volumes of the two cylinders are equal. Some of them explained that 'The volume of the two cylinders are the same because they are made from identical sheets of paper', or, similarly, that 'the volumes of the two cylinders are the same because the areas of the sheets are the same'. Another argument, based on compensation reasoning and more frequently mentioned by the older students, was: 'The volumes are the same as one cylinder is taller but thinner than the other'.

These results show that, indeed, students who conserved the area 'conserved' the volume as well. The similarity between the curves describing the development of the conservation of area and the 'conservation' of volume is striking. Moreover, Figure 1 shows that starting from grade 5, most children 'conserved' both area and volume and there is no evidence of a decrease with age in students' incorrect judgments of 'conservation' of volume.

These results support the hypothesis that children's realization of the qualitative and/or quantitative equality in quantity A ($A_1 = A_2$) in conservation tasks elicites the use of the rule 'Same A–same B'.

Evidence for the role of the intuitive rule 'Same A–same B' can be found in other cases related to mathematics and science. Here we provide two examples:

*Area and perimeter of geometrical shapes.* The mathematics education literature reports that many students and adults adhere to the view that shapes with the same perimeter must have the same area (e.g., Dembo, Levin and Siegler, 1997; Hirstein, 1981; Hoffer and Hoffer, 1992; Linchevsky, 1985; Shultz, Dover and Amsel, 1979; Ronen, 1996; Walter, 1970; Woodward and Byrd, 1983). These studies interpreted students' responses as resulting from a misunderstanding of the relationship between the concepts of area and perimeter, i.e., students were believed to think that 'shapes with the same perimeter must have the same area and vice versa'. Our claim is that this response could be viewed in a broader perspective, as resulting from an application of the intuitive rule 'Same A–same B (same perimeter–same area, same area–same perimeter; and also same object–same perimeter/area).

*Weight and volume of water.* Several studies in science education reported that students tend to confuse mass and volume. Meged (1978), who studied

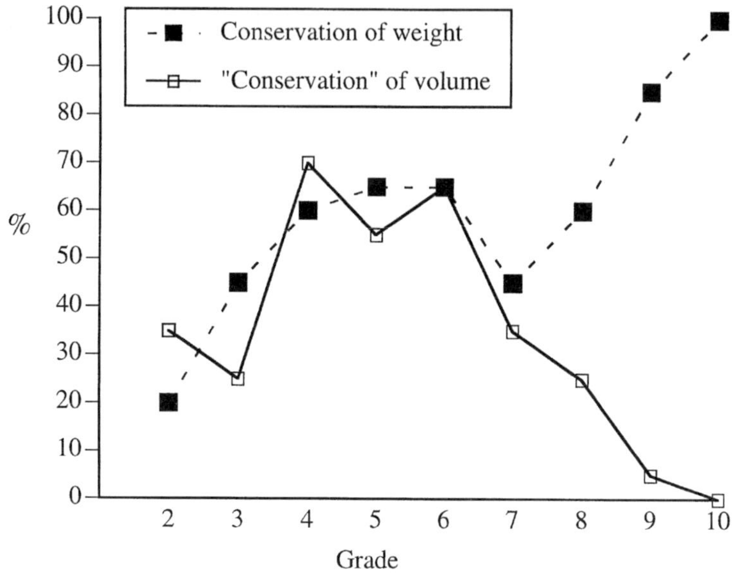


*Figure 2.* Distribution of equality judgment, by age, to the task of weight and volume of water

students' conceptions of density, presented 2nd to 9th graders with two vials containing equal amounts of water. Both vials were corked and a tube was inserted through the cork. One of the vials was heated, the water expanded, and consequently, its level rose in the tube. The children were asked to compare weight and the volume of water in the two vials before and after heating. In this case the weight of the water was conserved ($W_1 = W_2$) but the volume, after heating, was larger ($V_2 > V_1$).

It was found that children below grade 4 did not conserve the weight and claimed that the heated water weighed more than the unheated water because 'the level of water is higher'. (Similar results were reported by Piaget and Inhelder, 1974.) From grade 4 on, most students correctly judged that the weight of water in the two vials was the same, arguing that 'it's the same water, and it was only heated', or 'nothing was added or subtracted' (see Figure 2).

In respect to volume before and after heating, most 2nd and 3rd graders correctly judged that the volume of the heated water was larger, claiming that 'the level of water is higher'. Most 4th to 6th graders claimed that the volume of the heated water was equal to that of the unheated water, explaining that 'it's the same water, therefore it's the same volume'. In the upper grades there was an increase in correct judgments, accompanied by references to the particulate nature of matter. This increase in correct responses might be the result of instruction related to the particulate nature of matter, which probably constrains the use of the rule (children in Israel study the particulate nature of matter in grade 7). These data show that many students in grades 4 to 6 'conserved' both weight and volume. These high percentages of 'the same water–the same weight–the same volume' reasoning suggest once more the coercive effect of the intuitive rule 'Same A–same B' which in this case, until a certain age, overrules obvious perceptual input.

We have so far related to one logical scheme, namely conservation. We have shown that students who start conserving a certain quantity often 'conserved' another quantity which is in fact not conserved under the specific manipulation; their argument was that: 'Same A (the quantity conserved)–same B (a quantity not conserved)'.

Are there any other instances in which an acquired logical scheme, which leads to an equality judgment, is overgeneralized, leading to incorrect judgments of the type 'Same A–same B'? In other words, is the use of the rule 'Same A–same B' associated only with the conservation scheme or is it activated also when equality in quantity A ($A_1 = A_2$) is deduced through another logical scheme? We shall relate to the scheme of proportion.

### 3.2. *Proportion*

In the context of the development of probabilistic thinking, Schrage (1983) and Fischbein and Schnarch (1997) asked students to respond to the problem presented in Table II. The Table also presents the results reported by Fischbein and Schnarch.

According to the law of large numbers, as the sample size (or the number of trials) increases, the relative frequencies tend towards the theoretical probabilities. Consequently, the probability to get heads at least twice in tossing three coins is greater than that of getting at least 200 heads out of 300 tosses. However, as can be seen from Table II, a substantial number of students in each grade level argued that the probabilities are equal (and the frequency of this incorrect response increased with age). These students claimed that $\frac{2}{3} = \frac{200}{300}$ and therefore the probabilities are the same. This finding is explained, within the context of probability, as resulting from ignoring the role that sample size plays in calculating probabilities.

In terms of intuitive rules, we explain this behavior as follows: The equivalence in ratios, which is logically deduced through the proportion scheme, activates the intuitive rule 'Same A (proportion)–same B (probability)'. Fischbein and Schnarch similarly explained that:

TABLE II

Distribution of responses to comparison of probabilities (in %)

| The problem | Grades 5 | 7 | 9 | 11 |
|---|---|---|---|---|
| The likelihood of getting heads at least twice when tossing three coins is | | | | |
| smaller than | 5 | 5 | 25 | 10 |
| equal to | 30 | 45 | 60 | 75 |
| greater than | 35 | 30 | 10 | 5 |
| the likelihood of getting heads at least 200 times out of 300 times. | | | | |
| other answers | 5 | 10 | 0 | 0 |
| no answers | 25 | 10 | 5 | 10 |

The principle of equivalence of ratio imposes itself as relevant to the problem and thus dictates the answer. It is the evolution of this principle that shapes the evolution of the related misconception and causes it to become stronger as the student ages. (1997, p. 103).

The increase in the frequency of such responses follows the acquisition and stabilization of the proportion scheme.

This example confirms that the activation of the intuitive rule 'Same A–same B' is not limited to conservation tasks. The question whether there are other cases in which a logically deduced equality activates the use of the rule 'Same A–same B' is still open.

## 4. Discussion

Students' responses to a variety of mathematical and scientific comparison tasks were presented. In all these tasks, the two objects or systems to be compared were equal in respect to one quality or quantity ($A_1 = A_2$) but differnt in respect to another one ($B_1 \neq B_2$). In some of the tasks, the equality in quantity A was perceptually or directly given. In other cases, the equality in quantity A could be logically derived (through the schemes of conservation or proportionality). A common incorrect response to all these tasks, regardless of the content domain, was $B_1 = B_2$ because $A_1 = A_2$.

We regard all these responses as specific instances of the use of the intuitive rule 'Same A–same B'. This rule has the characteristics of an intuitive rule (Fischbein, 1987), as the response 'Same A–same B' seems *self evident* (subjects perceived statements they made on the basis of this rule as being true and in need of no further justification). This rule is used with great *confidence* and *perseverance* (often it persists in spite of formal learning that contradicts it). Moreover, the rule has attributes of *globality* (subjects tend to apply it to diverse situations), and *coerciveness* (alternatives are often excluded as unacceptable).

In the last decades attention has been paid, with reference to the development of specific mathematical and scientific concepts, to students' intuitive responses. Intuition is often described as a form of cognition related to specific content domain (concentration, area, perimeter, probability, volume, etc.) and students' responses are then explained in the context of this domain. Often, students' intuitive knowledge of certain concept or idea are not in line with the accepted scientific frameworks (Fischbein, 1987; Tversky and Kahneman, 1983). Based on the findings presented here, and in our other studies concerning intuitive rules, we suggest that students' responses to given mathematical and scientific tasks are often affected by common, external features of these tasks which trigger the use of these intuitive rules (Stavy and Tirosh, 1996; Tirosh and Stavy, 1996).

When one looks at students' responses to the various tasks presented in the paper, the rule 'Same A–same B' appears to be applied in a non-uniform way. In some tasks, only young children respond in accordance with it (e.g., length and distance, concentration and temperature). In other tasks, only older students and adults respond according to this rule (e.g., surface area and volume). Clearly, the application of 'Same A–same B' in the logically deduced tasks occurs only after the subjects acquired the relevant scheme (conservation and proportion). Another observation is that in some cases the incorrect application of the rule increases with age to reach a certain plateau (conservation and proportion), with no indication of decrement with age (e.g., surface area and volume), while in other cases, there are clear age-related indications of drop in application (e.g., weight and volume of water). Possibly, the response of students to a specific task is determined by the interaction of various factors including (1) different features of the task itself, such as perceptual and numerical characteristics of the objects to which the problem relates, and (2) solver-related characteristics such as age, instruction, repertoire of intuitive rules, logical schemes and formal knowledge.

Thus, the variation in students' responses to the weight and volume of water task, for instance, could be interpreted as follows: The obvious perceptual differences between the volumes of the heated and non-heated water support the use of the intuitive rule: 'More of A–more of B', which, in this case, leads to correct responses regarding the volumes of the water but to incorrect responses regarding the weight. Indeed, the vast majority of the young children correctly judged that the volume of the hot water is larger than that of the cold water. However older children who acquired conservation disregarded the obvious perceptual differences, and incorrectly argued, in line with the intuitive rule 'Same A–same B', that the volumes are equal. This response increased with the stabilization of the conservation scheme, and was frequent in the upper grades. Later on, acquired formal knowledge regarding the particulate structure of matter, along with the obvious perceptual differences, successfully competes with the intuitive rules. The ultimate result is an increase in the percentages of correct responses concerning both the volume and the weight.

We have suggested so far that 'Same A–same B' is an intuitive rule which could be activated by specific perceptual or logical input . What could be the origin of this intuitive rule? At this stage we can only suggest two speculative possibilities: (1) it is an innate, intuitive rule; the intuitive rule 'Same A–same B' may be one of a small set of universal, innate primitives; (2) it is an overgeneralization from successful experiences; often, both in everyday life and in school situations, the rule 'Same A–same B' is in fact applicable (e.g., 'same heights of juice in two identical cups–same amount to drink', 'same number of candies–same price'). It is reasonable to assume that children generalize such experiences into a universal maxim: 'Same A–same B'.

## 5. Educational Implications

It is widely accepted that knowledge about students' reasoning is crucial for teachers, policy makers and curriculum developers. Awareness of the role of intuitive rules in student responses to tasks which share specific, common external features can be applied for several purposes. Knowledge about intuitive rules has predictive power: it enables researchers, teachers and curriculum planners to foresee students' inappropriate reactions to specific situations, and this can help them plan appropriate sequences of instruction. We, for example, would suggest that instruction related to a certain topic begins with a situation in which the rule 'Same A–same B' is applicable. Later on related situations in which the rule does not apply could be presented. The differences between these two types of situations should be discussed, stressing the inapplicability of the intuitive rule in the second. This could help students to form the application boundaries of the intuitive rules. In addition we recommenf that students should be encouraged to criticize and test their own responses, relying on scientific, formal knowledge.

## REFERENCES

Cough, E.E. and Driver, R.: 1986, 'A study of consistency in the use of students' conceptual frameworks across different task contexts', *Science Education* 70(4), 473–496.

Dembo, Y., Levin, I. and Siegler, R.S.: 1997, 'A comparison of the geometric reasoning of students attending Israeli ultraorthodox and rain stream schools', *Developmental Psychology* 33(1), 92–103.

Erickson, G.L.: 1979, 'Children's conceptions of heat and temperature', *Science Education* 63(2), 221–230.

Fischbein, E.: 1987, *Intuition in science and mathematics: An educational approach*, D. Reidel, Dordrecht, The Netherlands.

Fischbein, E. and Schnarch, D.: 1997, 'The evolution with age of probabilistic, intuitively based misconceptions', *Journal for Research in Mathematics Education* 28(1), 96–105.

Hirstein, J.: 1981, 'The second area assessment in mathematics: Area and volume', *Mathematics Teacher* 74, 704–708.

Hoffer, A.R. and Hoffer, S.A.K.: 1992, 'Geometry and visual thinking', in T.R. Post, (ed.), *Teaching Mathematics in Grades K-8: Research-Based Methods* (2nd ed.), Allyn and Bacon, Boston.

Linchevsky, L.: 1985, 'The meaning attributed by elementary school teachers to terms they use in teaching mathematics and geometry', Unpublished doctoral dissertation, Hebrew University, Jerusalem, Israel (in Hebrew).

Livne, T.: 1996, 'Examination of high school students' difficulties in understanding the change in surface area, volume and surface area/ volume ratio with the change in size and/or shape of a body', Unpublished Master's thesis, Tel-Aviv University, Tel Aviv, Israel (in Hebrew).

Meged, H.: 1978, 'The development of the concept of density among children ages 6–16'. Unpublished Master's thesis, Tel-Aviv University, Tel Aviv, Israel (in Hebrew).

Noelting, G.: 1980a, 'The development of proportional reasoning and the ratio concept: Part I – differentiation of stages', *Educational Studies in Mathematics* 11, 217–253.

Noelting, G.: 1980b, 'The development of proportional reasoning and the ratio concept: Part II – problem structure at successive stages: problem solving strategies and the mechanism of adaptive restructuring', *Educational Studies in Mathematics* 11, 331–363.

Noss, R.: 1987, 'Children's learning of geometrical concepts through LOGO', *Journal for Research in Mathematics Education* 18(3), 343–362.

Nunes, T., Schliemann, A.D. and Carraher, D.W.: 1993, *Street Mathematics and School Mathematics*, Cambridge University Press, Cambridge.

Piaget, J. and Inhelder, B.: 1974, *The Child's Construction of Quantities*, Routledge & K. Paul, London.

Piaget, J., Inhelder, B. and Szeminska, A.: 1960, *The Child's Conception of Geometry*, Routledge & K. Paul, London.

Rojhany, L.: 1997, 'The use of the intuitive rule 'The more of A, the more of B': The case of comparison angles', Unpublished Master's thesis, Tel Aviv University, Tel Aviv, Israel (in Hebrew).

Ronen, E.: 1996, 'Overgeneralization of conservation', Unpublished Master's thesis, Tel Aviv University, Tel Aviv, Israel (in Hebrew).

Schrage, G.L.: 1983, '(Mis-)interpretation of stochastic models', in R. Scholz (ed.), *Decision Making Under Uncertainty*, North-Holland, Amsterdam, pp. 351–361.

Shultz, T., Dover, A. and Amsel, E.: 1979, 'The logical and empirical bases of conservation judgments', *Cognition* 7, 99–123.

Stavy, R. and Berkovitz, B.: 1980, 'Cognitive conflict as a basis for teaching quantitative aspects of the concept of temperature', *Science Education* 64, 679–692.

Stavy, R., Strauss, S., Orpaz, N. and Carmi, C.: 1982, 'U-shaped behavioral growth in ratio comparisons, or that's funny I would not have thought you were u-ish', in S. Strauss with R. Stavy (eds.), *U-Shaped Behavioral Growth*, Academic Press, New York, pp. 11–36.

Stavy, R. and Tirosh, D.: 1996, 'Intuitive rules in mathematics and science: The case of 'The more of A – the more of B', *International Journal of Science Education* 18(6), 653–667.

Strauss, S. and Stavy, R.: 1982, 'U-shaped behavioral growth: Implications for theories of development, in W.W. Hartup (ed.), *Review of Child Development Research*, University of Chicago Press, Chicago, pp. 547–599.

Tirosh, D.: 'Inconsistencies in students' mathematical constructs', *Focus on Learning Problems in Mathematics* 12(1), 111–129.

Tirosh, D. and Stavy, R.: 1996, 'Intuitive rules in science and mathematics: The case of 'Everything can be divided by two',' *International Journal of Science Education* 18(6), 669–683.

Tversky, A. and Kahneman, D.: 1983, 'Extensional versus intuitive reasoning: The conjunction fallacy in probability judgment, *Psychological Review* 90, 293–315.

Walter, N.: 1970, 'A common misconception about area', *Arithmetic Teacher* 17, 286–289.

Wiser, M. and Carey, S.: 1983, 'When heat and temperature were one', in D. Gentner and A.L. Stevens (eds.), *Mental Models*, Lawrence Erlbaum, Hillsdale, New Jersey, pp. 267–296.

Woodward, E. and Byrd, F.: 1983, 'Area: Included topic, neglected concept', *School Science and Mathematics* 83, 343–347.

DINA TIROSH and RUTH STAVY
*School of Education,*
*Tel Aviv University,*
*Tel Aviv 69978,*
*Israel*
*E-mail: Dina@post.tau.ac.il*

PAUL ERNEST

# FORMS OF KNOWLEDGE IN MATHEMATICS AND MATHEMATICS EDUCATION: PHILOSOPHICAL AND RHETORICAL PERSPECTIVES

ABSTRACT. New forms of mathematical knowledge are growing in importance for mathematics and education, including tacit knowledge; knowledge of particulars, language and rhetoric in mathematics. These developments also include a recognition of the philosophical import of the social context of mathematics, and are part of the diminished domination of the field by absolutist philosophies. From an epistemological perspective, all knowledge must have a warrant and it is argued in the paper that tacit knowledge is validated by public performance and demonstration. This enables a parallel to be drawn between the justification of knowledge, and the assessment of learning. An important factor in the warranting of knowledge is the means of communicating it convincingly in written form, i.e., the rhetoric of mathematics. Skemp's concept of 'logical understanding' anticipates the significance of tacit rhetorical knowledge in school mathematics. School mathematics has a range of rhetorical styles, and when one is used appropriately it indicates to the teacher the level of a student's understanding. The paper highlights the import of attending to rhetoric and the range of rhetorical styles in school mathematics, and the need for explicit instruction in the area.

## 1. Background

In the past decade or two, there have been a number of developments in the history, and philosophy and social studies of mathematics and science which have evoked or paralleled developments in mathematics (and science) education. I shall briefly mention three of these that have significance for the main theme of this paper, the import of rhetoric and justification in mathematics and mathematics education. Even though all of the developments I mention below are continuing sites of controversy, I merely list them rather than offer extended arguments in support of the associated claims, since this would draw me away from the main theme. Anyway such arguments can be found elsewhere (e.g., Ernest, 1997).

An important background development has been the emergence of fallibilist perspectives in the philosophy of mathematics. These views assert that the status of mathematical truth is determined, to some extent, relative to its contexts and is dependent, at least in part, on historical contingency. Thus a growing number of scholars question the universality,

*Educational Studies in Mathematics* **38:** 67–83, 1999.

absoluteness and perfectibility of mathematics and mathematical knowledge (Davis and Hersh, 1980; Ernest, 1997; Kitcher, 1984; Lakatos, 1976; Tymoczko, 1986). This is still controversial in mathematical and philosophical circles, although less so in education and in the social and human sciences. One consequence of this perspective is a re-examination of the role and purpose of proof in mathematics. Clearly proofs serve to warrant mathematical claims and theorems, but from a fallibilist perspective this warranting can no longer be taken as the provision of objective and ironclad demonstrations of absolute truth or logical validity. Mathematical proofs may be said to fulfil a variety of functions, including showing the links between different parts of knowledge (pedagogical), helping working mathematicians develop and extend knowledge (methodological), demonstrating the existence of mathematical objects (ontological), and persuading mathematicians of the validity of knowledge claims (epistemological), see, e.g., Hersh (1993) and Lakatos (1976). Below I elaborate further on the persuasive, epistemological role of proofs in mathematics.

The impact of these developments on education is indirect, as they do not lead to immediate logical implications for the teaching and learning of mathematics or the mathematics curriculum without the addition of further deep assumptions (Ernest, 1995). Nevertheless fallibilist philosophies of mathematics are central to a variety of theories of learning mathematics including radical constructivism (Glasersfeld, 1995), social constructivism (Ernest, 1991), and socio-cultural views (Lerman, 1994) which can have classroom consequences.

The second development is the emerging view that the social context and professional communities of mathematicians play a central role in the creation and justification of mathematical knowledge (Davis and Hersh, 1980; Kitcher, 1984; Latour, 1987). These communities are not merely accidental or contingent collections or organisations of persons incidental to mathematics. Rather they play an essential role in epistemology in two ways: their social organisation and structure is central to the mechanisms of mathematical knowledge generation and justification, and they are the repositories and sites of application and transmission of tacit and implicit knowledge (Ernest, 1997; Lave and Wenger, 1991; Restivo, 1992). In education, the vital roles played by the social and cultural contexts (Bauersfeld, 1992; Cobb, 1986, 1989), and the centrality of tacit and implicit knowledge in the mathematics classroom do not need to be argued, as they are already widely recognised (Bishop, 1988; Hiebert, 1986; Saxe, 1991; Tirosh, 1994).

Third, there is a move in the sociology and philosophy of science and mathematics to focus on communicative acts and performances of scientists and mathematicians, and in particular on their rhetorical practices (Fuller, 1993; Kitcher, 1991; Simons, 1989; Woolgar, 1988). In mathematics the parallel concern has been with writing genres and proof practices (Ernest, 1997; Livingston, 1986; Rotman, 1993). While there has been attention to the role of language in mathematics education for some time (Aiken, 1972; Austin and Howson, 1979; Durkin and Shire, 1991; Pimm, 1987; Skemp, 1982) it is only recently that an awareness of the significance of genres and rhetoric for the field are emerging (Ernest, 1993; Morgan, 1998; Mousley and Marks, 1991).

These background developments raise a number of issues concerning the form or forms of mathematical knowledge and the role and function of mathematical texts and proofs within the discipline itself and in the teaching and learning of mathematics. Whereas traditionally mathematical knowledge was understood as a collection of validated propositions, i.e., a set of theorems with proofs, a number of philosophers such as Ryle (1949), Polanyi (1959), Kuhn (1970) and Kitcher (1984) have argued that not all knowledge can be made explicit. The claim is that 'know how' and 'tacit' knowledge are important in all areas of human thought including mathematics. The argument for including tacit 'know how' as well as propositional knowledge as part of mathematical knowledge is that it takes human understanding, activity and experience to make or justify mathematics. Much that is accepted as a sign that persons are in possession of mathematical knowledge consists in their being able to carry out symbolic procedures or conceptual operations. To know the addition algorithm, proof by induction or definite integrals is to be able to carry out the operations involved, not merely to be able to state certain propositions. Thus what an individual knows in mathematics, in addition to publicly stateable propositional knowledge, includes mathematical 'know how'.

Kuhn (1970) argues that part of such knowledge in the empirical sciences consists of 'the concrete problem-solutions that students encounter from the start of their scientific education, whether in laboratories, on examinations, or at the ends of chapters in science texts. . . [and] technical problem-solutions found in the periodical literature.' (Kuhn, 1970, p. 187). Thus Kuhn claims that the experience of problem solving and of reading through various problem solutions leads to tacit knowledge of problem types, solution strategies, and acceptable modes of presentation of written work, i.e., tacit rhetorical norms (learned via instances). Kitcher extends the argument to mathematics and argues that both explicit propositional and tacit knowledge are important in mathematical practice, listing 'a language, a set of accepted statements, a set of accepted reasonings, a set of questions selected as important, and a set of meta-mathematical views (including standards for proof and definition and claims about the scope and structure of mathematics).' (Kitcher, 1984, p. 163). This list includes two knowledge components which are mainly tacit, namely language and symbolism, and meta-mathematical views, both of which have a strong bearing on the written, rhetorical aspects of mathematics.

The underlying language of mathematics is a mathematical sub-language of natural language (such as English or German) supplemented with specialised mathematical symbolism and meanings. It comes equipped with an extensive range of specialist linguistic objects, including mathematical symbols, notations, diagrams, terms, definitions, axioms, statements, analogies, problems, explanations, method applications, proofs, theories, texts, genres and rhetorical norms for presenting written mathematics. Mathematics could not be expressed without knowledge of its language, and most would argue more strongly that mathematics could not exist at all if mathematicians did not have knowledge of its language (Rotman, 1993; Thom, 1986). Although this knowledge includes explicit elements, as with any language, knowing how to use it is to a large extent tacit.

The set of meta-mathematical views includes a set of standards, the norms and criteria that the mathematical community expect proofs and definitions to satisfy if they are to be acceptable. Kitcher claims that it is not possible for the standards for proof and definition in mathematics to be made fully explicit. Exemplary problems, solutions, definitions and proofs serve as a central means of embodying and communicating the accepted norms and criteria. Like Kuhn, he argues that proof standards may be exemplified in texts taken as a paradigm for proof (as Euclid's Elements once did), rather than in explicit statements.

Thus mathematical knowledge not only encompasses a tacit dimension but also a concrete dimension, including knowledge of instances and exemplars of problems, situations, calculations, arguments, proofs, models, applications, and so on. This is not widely acknowledged, although the importance of knowledge of particulars has been recognised in a number of significant areas of research in mathematics education. For example, Schoenfeld (1985, 1992), in his research on mathematical problem solving, argues that experiences of past problems leads to an expanding knowledge-base which underpins successful problem solving. Current research on the situatedness of mathematical knowledge and learning also emphasises the role of particular and situational knowledge (Lave and Wenger, 1991; Saxe, 1991). More generally, in mathematics education the importance of implicit knowledge has been recognised for some time, and the categories of instrumental understanding (Mellin-Olsen, 1981; Skemp, 1976), procedural knowledge (Hiebert, 1986) and implicit knowledge (Tirosh, 1994) have been developed and elaborated to address it. These categories go beyond 'know-how', for as Fischbein (1994) argues, other forms of implicit knowledge such as tacit models are also important.

## 2. Mathematical knowledge and its justification

Drawing on Kitcher (1984) I have been proposing an extended concept of mathematical knowledge that includes implicit and particular components, but without reference to justification. Berg (1994) argues that this is illegitimate and that 'implicit knowledge' is a misnomer, for what passes under this name is either tacit belief (including misconceptions) or implicit method, since it lacks the robust justification that epistemologists require of knowledge. He is right that from an epistemological perspective knowledge only deserves its title if it has some adequate form of justification or warrant. However I reject his main critique because I believe that adequate warrants can be provided for tacit knowledge. Explicit knowledge in the form of a theorem, statement, principle or procedure typically has a mathematical proof or some other form of valid justificatory argument for a warrant. Of course the situation is different in mathematics education, as in other scientific, social or human science research, where an empirical warrant for knowledge is needed. Nevertheless, tacit knowledge can only be termed knowledge legitimately, in the strict epistemological sense, if it is justified or if there are other equivalent grounds for asserting it. However, since the knowledge is tacit, then so too its justification must be at least partly tacit, on pain of contradiction. So the validity of some tacit knowledge will be demonstrated implicitly, by the individual's successful participation in some social activity or form of life. Not in all cases, however, need the justification be tacit. For example, an individual's tacit knowledge of the English language is likely to be justified and validated by exemplary performance in conformity with the publicly accepted norms of correct grammar, meaning and language use, as related to the context of use. Thus a speaker's production of a sufficiently broad range of utterances appropriately in context can serve as a warrant for that speaker's knowledge of English.

This position fits with the view of knowledge in Wittgenstein's (1953) later work, according to which to know the meaning of a word or text is to be able to use it acceptably, i.e., to engage in the appropriate language games embedded in forms of life. Practical 'know how' is also validated by public performance and demonstration. Thus to know language is to be able to use it to communicate (Hamlyn, 1978). As Ayer says 'To have knowledge is to have the power to give a successful performance' (Ayer,

1956, p. 9). Such a validation is to all intents and purposes equivalent to the testing of scientific theories in terms of their predictions. It is an empirical, predictive warrant. It is a weaker warrant than a mathematical proof, for no finite number of performances can exhaust all possible outcomes of tacit knowledge as a disposition (Ryle, 1949), just as no finite number of observations can ever exhaust the observational content of a scientific law or theory (Popper, 1959). Thus tacit knowledge of mathematics can be defended as warranted knowledge provided that it is supported by some form of justification which is evident to a judge of competence. On this basis, what an individual knows in mathematics, in addition to publicly stateable propositional knowledge (provided it is warranted), includes her tacit knowledge. However what is warranted in this case is not the tacit knowledge, but the individual as possessor of that knowledge. We can then assert that Gerhard knows German, or Alicia knows proof by mathematical induction. So the tacit knowledge of an individual can be warranted as the knowledge of that individual.

There is a strong analogy between this warranting of an individual's tacit knowledge and the practices of assessment of knowledge and understanding in mathematics education and in education generally. To demonstrate knowledge of mathematics, within the institutionalised settings of mathematics assessment, requires successful performance at representative mathematical tasks. Of course there are technical and pragmatic issues involved in educational assessment. These include strategies for selecting mathematical tasks to be representative of the curriculum, i.e., the conventionally determined selection of mathematical knowledge in question. Other techniques are deployed so that a person's performances on the tasks selected are both valid and reliable predictors of the targeted skills and capacities, including the ability to reproduce and apply knowledge.

Thus just as there is a range of different types of knowledge of mathematics which is warranted in different ways, so too there is a range of different types of knowledge, each with associated means of warranting them, in mathematics education. However, a complication arises in mathematics education in that a fully explicit statement of an item of propositional knowledge can provide evidence of personal knowledge at a number of different levels. The bare recall of explicit verbal statements is typically placed at the lowest cognitive level of educational taxonomies of knowledge, such as those of Avital and Shettleworth (1968) and Bloom et al. (1956). To demonstrate that verbal statements are a part of warranted personal knowledge, as opposed to personal belief or acquaintance with others' knowledge, necessitates the knower demonstrating her possession of a warrant for that knowledge, typically a proof in mathematics. Thus to be able to produce a warrant for an item of knowledge, and to explain why it is a satisfactory warrant, is a higher level skill, often corresponding to the highest cognitive level in Bloom's taxonomy, the level of evaluation. (It may be observed that evaluation skills are to a significant extent implicit.) However, to simply recall a proof learnt by heart once again corresponds to the level of recall, illustrating that it is difficult to judge the cognitive level of a person's performance in mathematics without contextual information.

In addition to the highly rated implicit skills deployed in evaluation, tacit knowledge demonstrated in terms of being able to apply known methods, skills or capacities strategically to unfamiliar problems is also highly rated in terms of cognitive level. In Bloom's taxonomy this typically corresponds to the levels of analysis and synthesis. Although this particular taxonomy is now regarded as dated, its hierarchy of cognitive levels corresponds in gross terms to most recent assessment frameworks such as those in National Council of Teachers of Mathematics (1989) and Robitaille and Travers (1992). In Avital and Shettleworth's (1968) mathematics specific taxonomy, problem solving itself constitutes the highest level.

At the highest cognitive level of Bloom's taxonomy is evaluation; the ability to critically evaluate the knowledge productions of self and others. Whilst such productions must be based in concrete representations, such as the answers or solutions to problems, projects, reports, displays, models, multi-media presentations, or even performances, they may reflect the deployment of knowledge of all types and levels, including explicit and tacit knowledge. Such evaluation draws upon meta-mathematical knowledge of standards of proof, definition, reasoning, presentation and so on, knowledge which is primarily tacit, as is much of the knowledge deployed by experts in any field (Dreyfus and Dreyfus, 1986). Expert evaluative thought is, for example, a necessary skill for the teacher of mathematics, in order to make assessments of student learning. It is also a necessary skill for the research mathematician, not only in order to judge the mathematical knowledge productions of others, but as a skill that the mathematician must internalise and apply to her own knowledge productions, as an inner self-critical faculty.

## 3. THE ACCEPTANCE OF KNOWLEDGE AND RHETORIC

Although traditionally it has been thought that the acceptance of mathematical knowledge depends on having a logically correct proof, there is growing recognition that proofs do not follow the explicit rules of mathematical logic, and that acceptance is instead a fundamentally social act (Kitcher, 1984; Lakatos, 1976; Tymoczko, 1986; Wilder, 1981). 'A proof becomes a proof after the social act of "accepting it as a proof". This is as true of mathematics as it is of physics, linguistics and biology.' (Manin, 1977, p. 48). From such a social perspective the structure of a mathematical proof is a means to its epistemological end of providing a persuasive justification, a warrant for a mathematical proposition. To fulfil this function, a mathematical proof must satisfy the appropriate community, namely mathematicians, that it follows the currently accepted adequacy criteria for a mathematical proof. But these criteria are largely tacit, as every attempt to formalise mathematical logic or proof theory explicitly has failed to capture mathematicians' proof practices (Davis and Hersh, 1980; Ernest, 1991, 1997; Lakatos, 1976; Tymoczko, 1986).

> If, however, we think of 'rational certainty' as a matter of victory in argument rather than of relation to an object known, we shall look toward our interlocutors rather than to our faculties for the explanation of the phenomenon. If we think of our certainty about the Pythagorean Theorem as our confidence, based on experience with arguments on such matters, that nobody will find an objection to the premises from which we infer it, then we shall not seek to explain it by the relation of reason to triangularity. Our certainty will be a matter of conversation between persons, rather than an interaction with nonhuman reality. (Rorty, 1979, pp. 156–157)

As Rorty indicates, the deployment of informed professional judgement based on tacit knowledge (coupled with the persuasive power of the warrant) is what underpins the acceptance of new mathematical knowledge, not the satisfaction of explicit logical rules or correspondance with 'mathematical reality'. Likewise in mathematics education, the teacher's decision to accept mathematical answers in a student's work depends in part on the teacher's professional judgement, not exclusively on fixed rules of what is correct and incorrect. Teachers' views of correctness do play an important part in their judgements, but so do their aims and intentions relative to the given educational context. For example, despite its mathematical correctness, a pupil's answer of $\frac{1}{4} + \frac{1}{4} = \frac{2}{4}$ may be marked as wrong when the teacher desires the answer to be given in lowest form (i.e., $\frac{1}{2}$). As in the case of new mathematical knowledge productions, such judgements relate to the shared criteria, practices, and context and culture of the mathematics education and mathematics communities.

It may be thought that a teacher's judgement of correctness is a very local and subjective thing, compared with the verdict of the mathematical community on a new would-be item of mathematical knowledge. This is true, as it might be for a fellow mathematician's on the spot view of a new mathematical proof. However the proper comparison for the warranting mechanisms in research mathematics is with the educational institutions of assessment, with their rigorous protocols for examination procedures, marking, and external moderation and scrutiny. It is these institutions which certify individuals' mathematical knowledge, and which provide the proper analogue of the mathematical community's warranting procedures, especially when these are seen from a fallibilist perspective as essentially located in social institutions. Putting the institutional issues aside, the crucial issue in the present context concerns the criteria involved in professional judgements in both the mathematical research and education communities. My claim is that there are a variety of rhetorical styles that novices (both researchers and learners) are expected to master, in addition to other areas of knowledge and expertise.

As mentioned above, recently a rhetoric of the sciences movement has emerged in the philosophy and sociology of science. This is primarily concerned to acknowledge and describe the stylistic forms used by scientists to persuade others of the validity of their knowledge claims (Fuller, 1993; Latour, 1987; Nelson et al., 1987; Simons, 1989; Woolgar, 1988). Instead of being used pejoratively, the word 'rhetoric' is used to by these scholars to indicate that style is inseparable from content in scientific texts, and is equally important. 'Scholarship uses argument, and argument uses rhetoric. The "rhetoric" is not mere ornament or manipulation or trickery. It is rhetoric in the ancient sense of persuasive discourse. In matters from mathematical proof to literary criticism, scholars write rhetorically.' (Nelson et al., 1987, pp. 3–4). At base, 'rhetoric is about persuasion' (Simons, 1989, p. 2), and logic and proof provide the strongest rational means of persuasion available to humankind. As in the other sciences, the rhetoric of mathematics plays an essential role in maintaining its epistemological claims (Rotman, 1988, 1993). 'Even in the most austere case, namely mathematics, a rhetorical function is served by the presentation of the proof.' (Kitcher, 1991, p. 5).

Thus my claim is that rhetorical form plays an essential part in the expression and acceptance of all mathematical knowledge (Ernest, 1997). However, to persuade mathematical critics is not to fool them into accepting unworthy mathematical knowledge; it is to convince them that the actual proofs tendered in mathematical practice are worthy. Both the content and style of texts play a key role in the warranting of mathematical knowledge, and both are judged with reference to the judge's experience of a mathematical tradition, and the associated tacit knowledge, rather than with reference to any specific explicit criteria.

In fact, there are varying accepted rhetorical styles for different mathematical communities and subspecialisms. Knuth (1985) compared an arbitrarily chosen page (page 100) in nine mathematical texts from different subspecialisms and found very significant differences in style and content. This supports the claim that there is no uniform style for research mathematics and that no wholly uniform criteria for the acceptance of mathematical knowledge exist, since different subspecialisms have varying rhetorical (and contentual) requirements. This is confirmed by a more recent study by Burton and Morgan (1998). They analysed 53 published research papers in mathematics and the first author interviewed the mathematicians who had written them. The study found substantial variations in writing styles and rhetoric within and across mathematical specialisms. Furthermore, interviewees explicitly acknowledged that writing style as well as content quality played a key role in journal editors' and reviewers' responses to their submitted papers. As one interviewee said: 'you learn that you certainly do have to write . . . in exactly the form the editor wants or else you won't get to referee those papers [in your specialism] and they won't referee yours.' (p. 3)

Elsewhere I have specified in outline some of the general stylistic criteria that mathematical knowledge representations or texts are required to satisfy within the research mathematics community (Ernest, 1997). Bearing in mind that there are different genres of mathematical writing in schools too (Chapman, 1995; Mousley and Marks, 1991), some of the criteria of rhetorical style which generally apply to school mathematics are as follows. In order to be acceptable, a mathematical text should:

- Use a restricted technical language and standard notation
- Use spare, minimal overall forms of expression.
- Use certain accepted forms of spatial organisation of symbols, figures and text on the page
- Avoid deixis (pronouns or spatio-temporal locators).
- Employ standard methods of computation, transformation or proof (Ernest, 1997).

These criteria are of course far from arbitrary. They depersonalise, objectify and standardise the discourse, and focus on the abstract and linguistic objects of mathematics alone. They serve an important epistemological function, both in delimiting the subject matter, and simultaneously persuading the reader that what is said is appropriately standard and objective. Thus the rhetorical style demands on learner-produced texts concern an elementary and partial justification of the answers derived in tasks. They provide evidence for the teacher that the intended processes and concepts are being applied. However, there are significant variations in the rhetorical demands of teachers in different contexts, indicating that they are to a greater or lesser extent conventional. There are also variations of genre within the mathematics classroom, which bring variations in rhetorical demand with them (Chapman, 1997).

Thus, for example, a traditional or 'standard' school mathematics task as presented by a teacher is represented textually or symbolically, specifying a starting point, and indicating a general goal state, i.e., answer type. Thus a completed mathematical task recorded by a learner on paper is either an elaborated single piece of text (e.g., a 3 digit column addition), or a sequence of distinct inscriptions (e.g., the solution of a quadratic equation). In each case, carrying out the task usually involves a sequence of transformations of text, employing approved procedures. In addition to the required goal, i.e., the 'answer', the rhetorical mode of representation of these transformations is the major focus for negotiation between learner and teacher. Thus in the case of 3 digit column addition, the learner will commonly be expected to write the 'sum' on 3 lines, with one or two horizontal lines separating figures, the digits in vertical columns, and to indicate any units regrouped as tens. In more extended, sequentially represented tasks, the learner will be expected to use standard transformations, to represent the intermediate steps in conformity with accepted practice, and will often be expected to label the final answer as such (Ernest, 1993).

This last reference also points out the disparity between the learner's processes in carrying out a mathematical task, and its representation as a text. The text produced as answer to a mathematics task is a 'rational reconstruction' (Lakatos, 1978) of the derivation of the answer, and does not usually match exactly the learner's processes in deriving the answer. The disparity is usually determined by the rhetorical demands of the context, i.e., what are accepted as the standard means of representing procedures and tasks. This phenomenon is better recognised in school science, where there is widespread acknowledgement that student records of experiments are not personal accounts, but conform to an objectivised rhetoric which requires headings such as 'apparatus', 'method', 'observations', 'results', etc., in an impersonal and strictly regulated style of account. This reflects the fact that 'scientific writing is a stripped-down, cool style that avoids ornamentation' (Firestone, 1987; 17). It serves to reinforce the widespread objectivist philosophical assumptions of science and scientific method (Atkinson, 1990; Woolgar, 1988).

In contrast to 'standard' tasks, the introduction of project or investigational work in school mathematics (i.e., in 'progressive' or inquiry mathematics teaching) usually involves a major shift in genre (Richards, 1991). For instead of representing only formal mathematical algorithms and procedures, with no trace of the authorial subject, the texts produced by the student may also describe the subjective judgements and thought processes of the learner, as well as their justification. This represents a major shift in genre and rhetorical demand away from an impersonal, standard code towards a more personal account of mathematical investigation. There are often difficulties associated with such a shift related to the trained 'standard' expectations of pupils, parents, administrators and examination bodies (Ernest, 1991, 1998). Morgan (1998) illustrates some of these difficulties in her valuable study of teachers' expectations and responses to tasks of this type, such as their desire for standard terminology.

One of the few mathematics educators who acknowledges the importance of rhetorical knowledge in mathematics is Skemp (1979), although he uses a different terminology. Skemp distinguishes three types of understanding in learning mathematics: instrumental, relational and logical understanding. The distinction between instrumental and relational understanding due to Skemp (1976) and Mellin-Olsen (1981), referred to above, is well known. Instrumental understanding can be glossed in terms of tacit content knowledge of the methods of mathematics, i.e., knowing how to perform the methods and procedures to complete a task. Relational understanding can be partly glossed in terms of explicit content knowledge, but it also involves understanding the justification of the content, i.e., knowing both how to complete a task and why the approach works. Thus it relates to understanding the relationship between the task and content and a larger matrix of mathematical knowledge, and it requires the ability to offer an explicit explanation or justification.

The third element, logical understanding, encompasses both instrumental and relational understanding but goes beyond them, also including tacit knowledge of form (i.e., rhetorical knowledge). It involves knowing how to perform a mathematical task, knowing why the method works (i.e., being able to justify it verbally) and being able to express the working and solution of the task 'correctly' in written or symbolic form, i.e., having mastery of the rhetorical demands of school mathematics in the appropriate context. The inclusion of logical understanding is an under-recognised innovation of Skemp's. Skemp himself did not acknowledge the possibility of different standards of 'correctness' or different context-bound rhetorical demands for school mathematics, and probably accepted that a unique all-encompassing set of standards of 'correctness' exists, albeit localisable differently according to the educational context. Nevertheless Skemp had the prescience to acknowledge the difference between knowing why a procedure or task solution method works (e.g., being able to justify it informally or verbally) and being able to express the working and solution of the task in standard written form. This difference includes knowledge of the conventions and rhetorical demands of written school mathematics.

The rhetoric of school mathematics is important and deserves increased attention, for a number of reasons. First of all, there is the growing acceptance in mathematics and mathematics education circles of fallibilist and social philosophies of mathematics which point up the import of the discourse and rhetoric of mathematics, just as has been happening in science and science education. If the discourse of mathematics is no longer seen as purely logical, as an inevitable consequence of the discipline, but as having a contingent persuasive function varying with context, then the rhetoric of mathematics must be explicitly addressed and taught in the lecture hall and classroom. Since, on this reading, it is largely conventional, learners cannot be expected to learn it without explicit instruction.

Second, there is a need for a shift away from an overemphasis on students' subjective conceptions and thought processes, which is sometimes associated with constructivist views of learning. According to these perspectives, learning consists of the elaboration of subjective knowledge structures in the learner's mind, and the acquisition and elaboration these is primary, whereas public mathematical activities such as working written mathematical tasks or assessment exercises is secondary. Such a view separates the context of acquisition of knowledge from the context of its assessment or justification, and prioritises the former at the expense of the latter. This is problematic because without teacher or peer correction, i.e., formal or informal assessment and feedback, learners will not have their conceptions and actions 'shaped', and cannot know that they are mastering the intended mathematical content correctly. Quine (1960, pp. 5–6) stresses this need in general terms. 'Society, acting solely on overt manifestations, has been able to train the individual to say the socially proper thing in response even to socially undetectable stimulations.' Thus the individual construction of knowledge must be complemented by public interaction and response, both corrective and corroborative. Quine goes on to elaborate this with a simile.

> Different persons growing up in the same language are like different bushes trimmed and trained to take the shape of identical elephants. The anatomical details of twigs and branches will fulfil the elephantine form differently from bush to bush, but the overall outward results are alike. (Quine, 1960, p. 8)

Likewise, learners may construct individual and sometimes idiosyncratic personal understandings of mathematics but effective teaching must shape their mathematical performances and representations. Learning to shape one's own mathematical representations involves engagement with the rhetoric of mathematics, which is thus central to both the context of learning and the context of instruction and assessment.

Thus the rhetoric of school mathematics helps overcome the false dichotomy between learning and instruction/assessment. This parallels the current challenge to the absolutist dichotomy between the contexts of discovery and justification in philosophy (Popper, 1959; Reichenbach, 1951). The challenge is being mounted by modern fallibilists who increasingly locate the rhetoric of the sciences at the intersection of the contexts of discovery and justification (Ernest, 1997). Interestingly, traditional rhetoric dating back to the times of Aristotle and the Port Royal logicians includes the subdivisions of invention and instruction, anticipating the distinction contexts of discovery and justification, but without assuming their disjointedness (Fuller, 1993; Leechman, 1864). This traditional rhetorical distinction anticipates even better the parallel drawn here with the contexts of learning and instruction/assessment.

From a Vygotskian or social constructivist perspective, aiding and guiding the learner to develop her powers of written mathematical expression, i.e., mathematical rhetoric, is an essential activity for the teacher or informal instructor, in the zone of proximal development. For only under explicit guidance can the learner master, internalise and appropriate this rhetorical knowledge, in a piecemeal fashion.

In conclusion, it can be said that an epistemological perspective on mathematics and school mathematical knowledge foregrounds assessment and the warranting of knowledge. Both mathematical knowledge and mathematical knowers are judged within social institutions, those of research mathematics and mathematics schooling/assessment, respectively. Only explicit knowledge is directly warranted in the former context, although a wide range of types of tacit knowledge plays a part there both in the invention and warranting of mathematical knowledge (Ernest, 1997). In the contexts of schooling and educational assessment individuals' grasp of all types of knowledge is both developed and warranted. Since it plays an important part in both the development and assessment of learning I suggest the rhetoric of school mathematics needs increased attention by mathematics educators, as is currently happening with the rhetoric of the sciences. Needless to say, this is not proposed as an alternative to the development of understanding and capability in mathematics, but as a complementary and hitherto neglected element of these capacities.

## REFERENCES

Aiken, L. R.: 1972, 'Language factors in learning mathematics', *Review of Educational Research* 42(3).

Atkinson, P.: 1990, *The Ethnographic Imagination: Textual Constructions of Reality*, Routledge, London.

Austin, J. L. and Howson, A. G.: 1979, 'Language and mathematical education', *Educational Studies in Mathematics* 10, 161–197.

Avital, S. M. and Shettleworth, S. J.: 1968, *Objectives for Mathematics Learning*, Ontario Institute for Studies in Education, Toronto, Canada.

Ayer, A. J.: 1956, *The Problem of Knowledge*, Penguin Books, London.

Bauersfeld, H.: 1992, 'Classroom cultures from a social constructivist's perspective', *Educational Studies in Mathematics* 23, 467–481.

Berg, J.: 1994, 'Philosophical remarks on implicit knowledge and educational theory', in D. Tirosh, *Implicit and Explicit knowledge: An Educational Approach*, Ablex Publishing Co., Norwood, NJ, pp. 245–253.

Bishop, A. J.: 1988, *Mathematical Enculturation*, Kluwer, Dordrecht, The Netherlands.

Bloom, B. S. (ed.): 1956, *Taxonomy of Educational Objectives, Part 1: The Cognitive Domain*, David McKay, New York.

Burton, L. and Morgan, C.: 1998, 'Mathematicians writing', unpublished paper.

Chapman, A.: 1995, 'Intertextuality in school mathematics: The case of functions', *Linguistics and Education* 7, 243–262.

Chapman, A.: 1997, 'Towards a model of language shifts in mathematics learning', Paper presented at British Society for Research into Learning Mathematics Conference, University of Oxford.

Cobb, P.: 1986, 'Contexts, goals, beliefs, and learning mathematics', *For the Learning of Mathematics* 6(2), 2–9.

Cobb, P.: 1989, 'Experiential, Cognitive, and anthropological perspectives in mathematics education', *For the Learning of Mathematics* 9(2), 32–42.

Davis, P. J. and Hersh, R.: 1980, *The Mathematical Experience*, Birkhauser, Boston.

Dreyfus, H. L. and Dreyfus, S. E.: 1986, 'Putting computers in their place', *Social Research* 53(1), 57–76.

Durkin, K. and Shire, B., (eds.): 1991, *Language in Mathematical Education. Research and Practice*, The Open University Press, Milton Keynes.

Ernest, P.: 1984, 'CXC mathematics: A caribbean innovation in assessment', *Educational Studies in Mathematics* 15, 397–412.

Ernest, P.: 1991, *The Philosophy of Mathematics Education*, Falmer Press, London.

Ernest, P.: 1993, 'Mathematical activity and rhetoric: Towards a social constructivist account', in N. Nohda (ed.), *Proceedings of 17th International Conference on the Psychology of Mathematics Education*, University of Tsukuba, Tsukuba, Japan.

Ernest, P.: 1995, 'Values, gender and images of mathematics: A philosophical perspective', *International Journal for Mathematical Education in Science and Technology* 26(3), 449–462.

Ernest, P.: 1997, *Social Constructivism as a Philosophy of Mathematics*, SUNY Press, Albany, New York.

Ernest, P.: 1998, 'The relation between personal and public knowledge from an epistemological perspective', in F. Seeger, J. Voight and U. Waschescio (eds.), *The Culture of the Mathematics Classroom*, Cambridge University Press, Cambridge, pp. 245–268.

Firestone, W.: 1987, 'Meaning in method: The rhetoric of quantitative and qualitative research', *Educational Researcher* 16(7), 16–20.

Fischbein, E.: 1994, 'Tacit Models', in D. Tirosh, Implicit and Explicit knowledge: An Educational Approach, Ablex Publishing Co., Norwood, NJ, pp. 96–110.

Fuller, S.: 1993, *Philosophy, Rhetoric and the End of Knowledge*, University of Wisconsin Press, Madison, WI.

Glasersfeld, E. von: 1995, *Radical Constructivism: A Way of Knowing and Learning*, Falmer Press, London.

Hamlyn, D. W.: 1978, *Experience and the Growth of Understanding*, Routledge and Kegan Paul, London.

Hersh, R.: 1993, 'Proving is Convincing and Explaining', *Educational Studies in Mathematics* 24(4), 389–399.

Hiebert, J. (ed.): 1986, *Conceptual and Procedural Knowledge: The Case of Mathematics*, Hillsdale, New Jersey.

Kitcher, P.: 1984, *The Nature of Mathematical Knowledge*, Oxford University Press, Oxford.

Kitcher, P.: 1991; 'Persuasion', in M. Pera, W. R. Shea (eds.), *Persuading Science: The Art of Scientific Rhetoric*, Science History Publications, New York, pp. 3–27.

Knuth, D. E.: 1985, 'Algorithmic Thinking and Mathematical Thinking', *American Mathematical Monthly* 92, 170–181.

Kuhn, T. S.: 1970, *The Structure of Scientific Revolutions*, 2nd ed., Chicago University Press, Chicago.

Lakatos, I.: 1976, *Proofs and Refutations*, Cambridge University Press, Cambridge.

Lakatos, I.: 1978, *The Methodology of Scientific Research Programmes* (Philosophical Papers Volume 1), Cambridge University Press, Cambridge.

Latour, B.: 1987, *Science in Action*, Harvard University Press, Cambridge, MA.

Lave, J. and Wenger, E.: 1991, *Situated Learning: Legitimate Peripheral Participation*, Cambridge University Press, Cambridge.

Leechman, J.: 1864, *Logic: Designed as a Introduction to the Study of Reasoning*, William Allan and Company, London.

Lerman, S. (ed.): 1994, *Cultural Perspectives on the Mathematics Classroom*, Kluwer, Dordrecht, The Netherlands.

Livingston, E.: 1986, *The Ethnomethodological Foundations of Mathematics*, Routledge and Kegan Paul, London.

Manin, Y. I.: 1977, *A Course in Mathematical Logic*, Springer Verlag, New York, p. 48.

Mellin-Olsen, S.: 1981, 'Instrumentalism as an Educational Concept', *Educational Studies in Mathematics* 12, 351–367.

Morgan, C.: 1998, *Writing Mathematically: The Discourse of Investigation*, The Falmer Press, London.

Mousley, J. and Marks, G.: 1991, *Discourses in Mathematics*, Deakin University Press, Geelong, Victoria, Australia.

National Council of Teachers of Mathematics: 1989, *Curriculum and Evaluation Standards for School Mathematics*, National Council of Teachers of Mathematics, Reston, VA.

Nelson, J., Megill, A. and McCloskey, D. (eds.): 1987a, *The Rhetoric of the Human Sciences*, University of Wisconsin Press, Madison, WI.

Pimm D.: 1987, *Speaking Mathematically*, Routledge and Kegan Paul Ltd, London.

Polanyi, M.: 1958, *Personal Knowledge*, Routledge and Kegan Paul, London.

Popper, K.: 1959, *The Logic of Scientific Discovery*, Hutchinson, London.

Quine, W. V. O.: 1960, *Word and Object*, Massachusetts Institute of Technology Press, Cambridge, MA.

Reichenbach, H.: 1951, *The Rise of Scientific Philosophy*, University of California Press, Berkeley, California.

Restivo, S.: 1992, *Mathematics in Society and History*, Kluwer, Dordrecht, The Netherlands.

Richards, J.: 1991, 'Mathematical Discussions', in E. von Glasersfeld (ed.), *Radical Constructivism in Mathematics Education*, Kluwer, Dordrecht, The Netherlands, pp. 13–51.

Robitaille, D. F. and Travers, K. J.: 1992, 'International atudies of achievement in mathematics', in D. A. Grouws (ed.), *Handbook of Research on Mathematics Teaching and Learning*, MacMillan, New York, pp. 687–709.

Rorty, R.: 1979, *Philosophy and the Mirror of Nature*, Princeton University Press, Princeton, NJ.

Rotman, B.: 1988, 'Towards a semiotics of mathematics', *Semiotica* 72(1/2), 1–35.

Rotman, B.: 1993, *Ad Infinitum The Ghost in Turing's Machine: Taking God Out of Mathematics and Putting the Body Back in*, Stanford University Press, Stanford California.

Ryle, G.: 1949, *The Concept of Mind*, Hutchinson, London.

Saxe, G. B.: 1991, *Culture and Cognitive Development: Studies in Mathematical Understanding*, Erlbaum, Hillsdale, New Jersey.

Schoenfeld, A.: 1985, *Mathematical Problem Solving*, Academic Press, New York.

Schoenfeld, A.: 1992, 'Learning to think mathematically', in D. A. Grouws (ed.), (1992) *Handbook of Research on Mathematics Teaching and Learning*, MacMillan, New York, pp. 334–370.

Simons, H. (Ed.): 1989, *Rhetoric in the Human Sciences*, Sage, London.

Skemp, R. R.: 1976, 'Relational understanding and instrumental understanding', *Mathematics Teaching* 77, 20–26.

Skemp, R. R.: 1979, 'Goals of learning and qualities of understanding', *Mathematics Teaching* 88, 44–49.

Skemp, R. R. (ed.) 1982, 'Understanding The Symbolism of Mathematics', *Visible Language* 16(3).

Thom, R.: 1986, 'Mathematics', in T. A. Sebeok (ed.), *Encyclopedic Dictionary of Semiotics*, Vol. 1, Mouton de Gruyter, Berlin, pp. 487–494.

Tirosh, D. (ed.): 1994, *Implicit and Explicit Knowledge: An Educational Approach*, Ablex Publishing Co, Norwood, NJ.

Tymoczko, T. (ed.): 1986, *New Directions in the Philosophy of Mathematics*, Birkhauser, Boston.

Wilder, R. L.: 1981, *Mathematics as a Cultural System*, Pergamon Press, Oxford.

Wittgenstein, L.: 1953, *Philosophical Investigations* (translated by G. E. M. Anscombe), Basil Blackwell, Oxford.

Woolgar, S.: 1988, *Science: The Very Idea*, Ellis Horwood/Tavistock, London.

*School of Education,*
*University of Exeter,*
*Exeter, Devon EX1 2LU,*
*U.K.*
*E-mail: PErnest@ex.ac.uk*

TOMMY DREYFUS

# WHY JOHNNY CAN'T PROVE

*(with apologies to Morris Kline)*

ABSTRACT. The one sentence answer to the question in the title is that the ability to prove depends on forms of knowledge to which most students are rarely if ever exposed. The paper gives a more detailed analysis, drawing on research in mathematics education and classroom experiences.

## 1. Introduction

Recent changes in mathematics teaching at all levels include attempts to make learning experiences more cooperative, more conceptual and more connected. As a consequence, students are more and more frequently asked to explain their reasoning; for example, Silver (1994) suggested that written explanations should become a prevalent feature of school mathematics and predicted that 'unless and until solution explanations and interpretations become a regular item on the menu of instructional activities in mathematics classrooms, ... there can be little hope of substantially improving the poor mathematics performance of American students' (p. 315). On another continent and for a different student age group, the custom of scientific debate has been firmly established since 1984 as an opportunity for deep learning experiences in the framework of a large first year university mathematics course (Alibert and Thomas, 1991). Other examples will be referred to below.

Occasions for mathematics students to make their reasoning explicit may arise for a number of reasons: A student may want to convince a classmate of a guess or conjecture during a collaborative phase; another student may have asked for help; or the teacher may try to obtain clarification about students' thinking in order to help them, to assess their progress, or attempt to move them from a descriptive to a justificative mode of thinking about what they are doing (Margolinas, 1992). In these cases, the explanations students are asked to provide are thus arguments, possibly even proofs. This increased emphasis on explanation, argument and proof is consistent with the continued importance of proof in mathematics (Hanna, 1995; Dreyfus, in press).

*Educational Studies in Mathematics* **38:** 85–109, 1999.

The different situations mentioned above may require different kinds of explanations. This paper will focus mainly on written explanations given by college students as response to questions by a teacher or textbook, for example in homework assignments. Even within this limited framework, several questions immediately arise: As teachers and as educators, on what basis do we or do we not accept a student's explanation? Which arguments are acceptable to us under which circumstances, which are not, and why? What criteria do we use and what are these criteria based on? Do we expect a proof when we ask for a justification? And do we expect a proof when we ask students to 'explain why'? To what degree does an explanation need to convince? And if so, does it need to convince a mathematician, the teacher, fellow students?

These questions cannot be answered in general, outside of the framework of a particular curriculum or course. But often, the questions are not even asked; or if they are asked, they remain unanswered even where answers are feasible and essential. A first aim of this paper is therefore to provide appropriate background for analyzing these questions. The paper thus has a descriptive rather than a normative character. In this respect, it is different from Kline's (1973) book from which the title was adapted.

On the other hand, some changes in instruction may well be indicated: Sample explanations, even from the more successful students show that frequently their criteria for acceptable explanations appear to differ from their teachers' criteria. For example, students often provide chronological accounts of actions carried out rather than pointing out connections and implications. One might therefore add another set of questions: What do students consider a satisfactory explanation? What is the basis of their conception of a satisfactory explanation? And what is the origin of the differences between students' and teachers' conceptions? The second aim of this paper is to identify some of the reasons for students' limited conceptions of explanation and proof; in other words, I will attempt to clarify why students cannot be expected to have a mathematician's concept of proof, of its function and of its purpose.

In Section 2, some sample explanations will be exhibited; most have been given by students, and they are intended to show how difficult the task is, even for the more successful ones. This point will be strengthened in Section 3 by means of a review of research on proof and explanation at the college, high school and elementary school levels. In Section 4, the influence of typical high school and college mathematics experience on students' image of explanation and proof will be reviewed. In Section 5, the review of research will be continued with epistemological and cognitive work on the distinction between proof and explanation; the discussion will also be carried beyond mathematics education to the topic of proof in mathematics itself. And in the concluding Section 6, the material from all the previous sections will be pulled together into an appraisal of the difficulties students have with explanations and proofs.

## 2. Sample Explanations

Abstract characterizations of explanations will be discussed later (Section 5). Instead, we start with a number of sample explanations which will be used to raise some pertinent questions. Most of these examples have been collected, somewhat randomly, from students participating in introductory university level courses such as calculus or linear algebra. Moreover, the examples are answers to questions and problems on which the students demonstrated a certain proficiency and some understanding. This choice has been made in order to focus on the characteristics of the explanations rather than on misunderstandings of the questions or the concepts involved in answering them. My concern is thus not with the question 'Why were the students unable to give the correct answer?' but rather with the question 'Why were the students unable to give a decent explanation in spite of the fact that they seem to have a satisfactory understanding of the question and its answer (or of the problem and its solution)?' In this respect, the focus of the present paper is different from that of other recent work; for example, Ferrari (1997) analyzed similar questions answered by a similar student population but was interested mainly in the analysis of the students' wrong answers and the conceptual reasons for these wrong answers; Vinner (1997), on the other hand, analyzed ways in which students altogether avoid to conceptually deal with the questions which are presented to them, and exhibited reasons for such behavior.

Most of the following examples have been chosen from take-home assignments which first year undergraduate students at two universities in two different countries handed in after having been given about a week to prepare the assignment. Students were explicitly advised that explanations of what they did and why they did it were crucial and would account for a substantial part of the grade. I have chosen cases in which the answers led me to believe that the question was meaningful to the student, and the student showed a substantial understanding of the procedures and concepts needed to answer it. Apart from that, neither the examples nor the students are representative in any sense but have been chosen for illustrative purposes. Their choice, including the topics, the level of mathematics, and the level of students have been influenced by my own personal bias and experience.

Ferrari characterized his weakest students as 'unable to use words to express even elementary mathematical ideas and relationships' (loc. cit., p. 2–262). The use of language to express mathematical relationships is a crucial constituent of explanations. As a consequence, most of the following examples relate, in some way, to the use of language.

*Example 1.*
Determine whether the following statement is true or false, and explain:
If $\{ v_1, v_2, v_3, v_4 \}$ is linearly independent, then $\{ v_1, v_2, v_3 \}$ is also linearly independent.

RP: True because taking down a vector does not help linear dependence.

The use of 'taking down' rather than, say, 'omitting' points to a lack of linguistic ability; this impression is compounded by the use of the word 'help' rather than a less vague but presumably more complex term. But ignoring the purely linguistic inadequacies of the explanation, we can try to speculate on its mathematical adequacy.

Maybe RP thought as follows: 'I know that adding a vector to a given linearly independent set of vectors might produce a linearly dependent set; on the other hand, adding a vector to a linearly dependent set will not produce a linearly independent set. In other words, adding a vector to a set "helps" the linear dependence of the set. Thus, omitting a vector from a set does not "help" its linear dependence. Since I was instructed to state my reasoning concisely, I will only write down an abridged version of the last sentence.'

It should be noted that even this expanded (and invented) explanation has mathematical and logical problems. It is not sufficiently sharp, from the mathematical point of view. By means of the introduction of the term 'help' for 'might produce' a vagueness is introduced which can be interpreted as 'in some cases adding a vector will produce a linearly dependent set' or as 'in all cases adding a vector will produce a linearly dependent set'. Such vagueness may be due to lack of conceptual clarity or to lack of linguistic ability, or to a combination. Next, the expanded version proceeds to take the converse of one part of the preceding sentence. Since I (rather than the student) invented the sentence preceding the converse, we cannot know on what the student's claim was based. But we may go further and ask whether the expanded version, even after replacing the word 'help' by a less vague one would constitute a satisfactory explanation: Isn't it almost tautological to state that the claim is true because adding a vector to a linearly dependent set will always produce a linearly dependent set? Shouldn't the student have also explained, using the definition of linear dependence, why adding a vector to a linearly dependent set cannot produce a linearly independent set? How far back does an explanation have to go? How deep does it need to be in order to count as an explanation?

The above example is far from being a special case; in fact, it is rather typical. Similar analyses could be carried out for the following two examples:

*Example 2.*
Determine whether the following statement is true or false, and explain: If $v_1$, $v_2$, $v_3$, $v_4$ are in $R^4$ and it is known that $v_3 = 0$, then the set $\{ v_1, v_2, v_3, v_4 \}$ is linearly dependent.

RC: True, the nontrivial solution is possible because $v_3$ is equal to *0*.

*Example 3.*
Is the following statement true or false; justify your answer: A system of n equations in n unknowns has at most n solutions.

TA: False; if we will discover a consistent system with a free variable, there will be $\infty$ solutions.

The most generous evaluation a teacher of a first year linear algebra class would presumably give to these answers is that they include important elements of the required explanation but are not substantial enough. Students at this stage of their education appear to find it extremely difficult to distinguish conciseness from lack of substance. I regularly have students who complain about my requirements arguing that they should not be required to write text because they are taking a mathematics class rather than a literature class. There are two extreme cases of not sufficiently substan tial explanations. One is stating a tautology rather than an explanation by simply repeating the claim. The other is not giving an explanation at all but a computation, as in Example 4.

*Example 4.*
Are the columns of the matrix A linearly independent?

$$A = \begin{bmatrix} 3 & 4 & 9 \\ -2 & -7 & 7 \\ 1 & 2 & -2 \\ 0 & 2 & -6 \end{bmatrix}$$

$$AW: \begin{bmatrix} 3 & 4 & 9 \\ -2 & -7 & 7 \\ 1 & 2 & -2 \\ 0 & 2 & -6 \end{bmatrix} \sim \begin{bmatrix} 1 & 2 & -2 \\ 2 & -7 & 7 \\ 3 & 4 & 9 \\ 0 & 2 & -6 \end{bmatrix} \sim \begin{bmatrix} 1 & 2 & -2 \\ 0 & 1 & -1 \\ 0 & 0 & 13 \\ 0 & 0 & 4 \end{bmatrix} \sim \begin{bmatrix} 1 & 0 & 0 \\ 0 & 1 & -1 \\ 0 & 0 & 0 \\ 0 & 0 & 4 \end{bmatrix}$$

Although I have no specific information about this, it may well be that AW was one of the students complaining about my 'literacy' requirements. He reacted to the question by correctly carrying out a computation (reduction of the matrix) from whose result the answer to the question can be read off. He might even have read the answer off (the three column vectors are linearly independent) but he did not consider it necessary to leave a written record of this; nor did he consider it necessary to establish the connection between the computation and the question: Why and how can the linear independence of the column vectors be read off the reduced matrix? We don't know whether AW could have provided the why and how; we only know that he didn't – presumably because to him the computation constitutes the most important part of answering the question rather than the answer itself or the explanation justifying the procedure. It is not at all obvious that the same aspects of an answer (or solution) are considered important by the teachers and students of beginning university mathematics courses.

Above, explanations lacking in substance were considered. Occasionally, students exaggerate in the opposite direction and 'explain' by writing down whatever comes to their mind and might possibly be related to the question. This results in texts which include all the elements needed for the required explanation, and with redundant information added. The following is a relatively mild case.

*Example 5.*
Prove that the equation $x^3+9x^2+33x-8=0$ has exactly one real root.

AM: [Defines $f(x) = x^3+9x^2+33x-8$; differentiates f and shows that the derivative has no real roots. Then continues:] The fact that the derivative has no roots means that there are no critical points at which one has to check the behavior of the function.
$\lim_{x\to\infty} f(x) = \infty$. $\lim_{x\to-\infty} f(x) = -\infty$.
The derivative of the function is always positive. The function exists for all x. The function increases always and therefore, because it goes from $-\infty$ to $\infty$ there is only one real root.

By omitting the redundant part and reordering the rest, AM's argument can be made into:

AM': The function exists for all x. The derivative of the function is always positive. The function increases always; it goes from $-\infty$ ($\lim_{x\to-\infty} f(x) = -\infty$) to $\infty$ ($\lim_{x\to\infty} f(x) = \infty$). Therefore, there is only one real root.

This argument is quite acceptable, although far from perfect; for example, it states that the function exists for all x but omits to state that it is differentiable for all x; it also concludes that there is 'only one root', not making it completely clear that the intention is 'exactly one root'. The question arises why the argument AM is less acceptable than AM' as an explanation of the fact that the function has exactly one root. From the student's point of view, is the omitted part really redundant? After all, we did need the fact that there are no critical points, and critical points are points at which often the behavior of the function has to be checked; in the case at hand, if there was a critical point, it might be one at which the function is not differentiable, and AM has demonstrated consciousness of this possibility. Moreover, and again from the student's point of view, AM presents all arguments needed to draw the correct conclusion; why should it be so important that the arguments appear in one order rather than another? To the teacher, it might appear that AM was not quite able to correctly combine all the elements presented into a coherent proof; to AM, it might appear that she has provided the required proof and beyond, some elaboration which adds to the minimum which is necessary.

Uncertainty about how to handle redundancy might lead some students to add unnecessary, even superfluous elements as above; on the other hand, it might lead others to omit elements which are relevant and necessary for a complete argument. The next example is a case in point.

*Example 6.*
Show that if AB and BA are both defined, then AB and BA are square matrices.

DM: By definition, if A is an $m \times r$ matrix and B is an $r \times n$ matrix, then AB is an $m \times n$ matrix. BA would then be an $r \times r$ matrix which is a square.

As in the earlier examples, several different interpretations of DM's explanation could be given, ranging from a rather severe critique of the explanation's inadequacy as a mathematical proof to a rather positive evaluation of the correctness and the relevance of the elements which are presented, and related to each other. While it is not difficult to point out how the explanation could be improved, we do not have the means to find out why the student did not give a better explanation. The characteristics of this as well as earlier student explanations show that giving an argument or explanation is a very difficult undertaking for beginning undergraduates from at least two points of view: In most cases, they still lack the conceptual clarity to actively use the relevant concepts in a mathematical argument; and, more generally, they have had little opportunity to learn what are the characteristics of a mathematical explanation.

The two final examples show, that not only students have such problems.

*Example 7.*
The series $\Sigma \frac{1}{k(k-1)}$ converges because S(2) = 1/2, S(3) = 2/3, S(4) = 3/4, S(10) = 9/10, S(100) = 99/100, and thus S(n) = (n-1)/n. For infinitely large values of n, the partial sums S(n), S(n+1), ... differ from the limit S=1 and consequently among themselves by an infinitely small quantity.

Did the student compute S(100) or guess it inductively? What exactly does the word 'thus' mean. And what notion of convergence was used? While this example looks typical for a modern calculus student, its treatment of convergence is historically based on Cauchy's (1821) definition: 'When the values successively attributed to the same variable approach indefinitely a fixed value, eventually differing from it by as little as one could wish, that fixed value is called the limit of all the others.' Seen from today's vantage point, Cauchy's definition lacks conceptual clarity (Lakatos, 1978) and Cauchy was able to use it to prove the (incorrect) result that the limit of any sequence of continuous functions is continuous.

*Example 8.*
The following exchange took place within a six hour period in June 1997 on an electronic list whose topic is post-calculus mathematics teaching and whose participants are university level mathematics teachers and researchers:

DE: Here is a problem from Arnold (1991) I've just managed to solve. Let f(x,y) be a polynomial with real coefficients, such that f(x,y) > 0 for all (x,y) in the plane. Does f necessarily achieve its minimum?

AD: It is too easy. It can't be a polynomial of odd degree, otherwise it will take both signs. So it's even degree, and bounded below by 0. It must have a positive minimum at a critical point, which will be its absolute minimum, so, yes, it attains its minimum. Am I missing something here?

RR: No; but one does not even need critical point theorems. Once you know P takes positive values only, knowing it to be a polynomial tells you the leading coefficient is positive, because the term of highest (combined) degree outclasses all the others combined as |x|+|y| $\longrightarrow$ infinity. Let P(a,b)=M, then choose a box outside of which P(x,y) > M, the box also big enough to contain the point (a,b). (Here the positivity of the leading term and the triangle inequality turn up.) Then by continuity alone, P has a minimum somewhere in the closed box, and since its value is < M there it is an absolute minimum for the whole plane.

SL: I don't find RR's proof convincing because the crucial step is unclear. There is no one term of highest degree. Obviously, if one can show that the homogeneous part of highest degree is positive definite, the rest follows, just as RR says. AD's proof is also not convincing. Why does even degree imply (in a two-variable polynomial) the existence of a critical point?

DS: Does f necessarily achieve its minimum? No. Take $f(x,y) = x^2 + (xy-1)^2$. It is clear that f(a,b) > 0, and f(a,b) = 0 implies a = 0, ab = 1, which is impossible. However, $f(a,1/a) = a^2$ can be as small a positive number as desired.

How are we to evaluate the (mistaken) explanations by AD and RR? They use relevant and correct input, they are convincing, and they support what many list readers, including this author, clearly expected to be the correct answer. On the other hand, the problem cannot be considered very difficult. Many calculus teachers use examples like $g(x,y) = xy^2/(x^2+y^4)$ in order to show that the two variable case is much more complex than the one-variable case; the function g converges to zero along every straight line through the origin but does not converge to zero at the origin. This is fairly similar in spirit to DS's function which has a minimum on every line through the origin but not in the plane. How could professional mathematicians go so wrong? Are explanations so difficult to give and judge?

To conclude this section, the questions which were raised are collected for further reference:

- Which aspects of an answer (or solution) are considered most important: computation, statement of the answer, relationship between computation and answer, procedural or conceptual? Are the same aspects considered important by the teachers and by the students of beginning university mathematics courses? How are the students supposed to know what the teacher considers important?
- How difficult is it to give and judge mathematical explanations? Should we see our students' imperfect explanations in a rather forgiving light?
- How deep does an explanation have to be? Does it always have to go all the way back to the relevant definitions? Does it have to live up to the same stringent criteria as a proof, namely to use only definitions and previously proved statements?
- How important is the order of the reasons from which a conclusion is to be drawn?
- Is redundancy wrong? Explanations given in teaching situations are often redundant; lecturers tend to repeat statements many times over, giving different points of view, and connections to various related concepts. Why then should a student's explanation be free of such redundancies?
- How much accuracy is required from an explanation? Under what circumstances may students use vague terms such as 'does not help' (example 1)?
- What counts as tautological? What might sound tautological to the teacher might constitute a considerable conceptual step for the student because (s)he is less well versed in the subject matter.
- Can we tell whether a student's problem is linguistic rather than conceptual? How should we deal with linguistic problems?

The examples and questions of this section have been collected here somewhat informally because we lack a better research base on student explanations in undergraduate mathematics. There is a shortage of research data, and it is one aim of this paper to point out the need for such research. A first conclusion – as informal as the data on which it is based – is that the task of explaining is extremely difficult, even for reasonably proficient students who were accepted into a university and exhibit some understanding of the topic.

## 3. Research on students' conceptions of proof

Student answers were selected for presentation in Section 2 only if they showed some features which can be considered as justificative; moreover, some well known cases like proof by example were not even illustrated. Nevertheless, the variety of the answers is great and may show different stages of development. With this in mind, mathematics educators have attempted to classify students' developing notions of proof. Balacheff (1987), for example, distinguishes pragmatic proofs and intellectual proofs, subdividing each into several subclasses; and Harel and Sowder (1998) propose a large set of schemes intended to make a classification of college students' proof-like productions possible.

Our aim here is not to classify such productions, or follow their development but rather to identify some of the reasons for the fact that many students appear to have a very limited conception of proof. Indeed, research results on students' conceptions of proof are amazingly uniform; they show that most high school and college students don't know what a proof is nor what it is supposed to achieve. Even by the time they graduate from high school, most students have not been enculturated into the practice of proving, or even justifying the mathematical processes they use.

Fischbein (1982), for example, provided close to 400 high school students with a proof of the statement 'For every integer n, the number $E = n^3 - n$ is divisible by 6'. Although over 80% of the students affirmed that they had checked the proof and found it to be correct, less than 70% agreed that $E = n^3 - n$ will always be divisible by 6, less than 40% concluded that a purported counterexample must contain a mistake, and less than 30% agreed that there was no need for additional checks in order to decide on the truth or falsity of the statement. Fischbein concluded that less than 15% of the students really understood what a mathematical proof meant.

Coe and Ruthven (1994) found that when proof contexts are data-driven, and students are expected to form conjectures by generalization or counter example, then students' proof strategies are primarily empirical. It seems that in such a context students are willing to replace deductive argument by a sufficiently diverse set of instances.

Similarly, Finlow-Bates, Lerman and Morgan (1993) found that many first year undergraduates had difficulties following chains of reasoning, and judged mathematical arguments according to empirical or aesthetic rather than logical criteria.

Martin and Harel (1989) provided preservice elementary teachers with correct deductive, incorrect deductive, and inductive arguments for the same statements. Every inductive argument they presented was accepted as a valid mathematical proof by more than half of their students; the acceptance rates for the deductive arguments were not much higher than those for the inductive ones; and the false deductive proofs were accepted by close to half of the students.

Finally, Moore (1994) found that even apparently trivial proofs are often major challenges for undergraduate mathematics majors.

Is it, then, a fact that students cannot argue mathematically at all? This would be an unwarranted conclusion. In fact, all examples provided in Section 2, while being rather far from constituting rigorous proofs, contain clear seeds of mathematical argumentation and justification. Moreover, several studies carried out at the upper elementary level show that in suitable environments some students develop promising abilities.

Maher and Martino (1996) report a sequence of eleven events in the development of one elementary student's justificative arguments over a five year period. While most tasks given to the student required the classification and organization of data, the student progressively developed not only her ability to classify systematically but more significantly, her ability to accompany the classification by verbal argumentation showing, for example, that the classification is indeed complete. The authors conclude that the student's interest in justifying arose out of her idea that mathematics should make sense.

Zack (1997) analyzed the work of a team of fifth grade students who considered patterns in a counting problem. They used what they knew of the patterns to refute arguments by other teams. Zack found evidence of conjecture, refutation, generalization, and aspects of proving.

In teaching a fifth grade classroom, Lampert (1990) consciously and systematically initiated and supported social interactions appropriate to making mathematical arguments. As a result, her students began 'to make assertions that were based on their inductive observation of patterns and to move back and forth between these observations and deductive arguments about why the patterns would continue, even beyond the numbers they had tested' (p. 49). She concluded that classrooms can be led in such a way that 'in [students'] talk about mathematics, reasoning and mathematical argument – not the teacher or the textbook – are the primary source of an idea's legitimacy' (p. 34).

These reports appear to show that, in terms of deductive argumentation, fifth graders may show as much ability as college students. One has to keep in mind, though, that the elementary school children were observed in classes carefully planned and taught so as to support mathematical reasoning, argument and justification. Therefore, the studies only show that the transition to deductive reasoning is possible, not that it normally happens. And the studies at the high school and college level show that it often does not happen. Much of the remainder of this paper, in particular the next section, will discuss reasons why it does not.

At the most general level, the reason is obviously that most students never learned what counts as a mathematical argument. Although this sounds trivial, it isn't: Yackel and Cobb (1996) have coined the term 'sociomathematical norms' in order to discuss how environmental influences (teachers, classroom activity, ...) determine students' mathematical beliefs and activity in the framework of a class or course; for example, '... what counts as an acceptable mathematical explanation and justification is a sociomathematical norm' (p. 461). Yackel and Cobb show how, in a second grade classroom, teacher and students interactively and consciously constitute sociomathematical norms regulating mathematical argumentation.

College textbook authors and teachers are rarely conscious of the need to establish sociomathematical norms, and their actions are often more apt to confuse rather than help students. College students do not usually read mathematics research papers, or see research mathematicians in action. But they do listen to lectures and participate in exercise sessions; they see and experience the talk and actions by their teachers; they read textbooks; they hand in assignments and tests, and they consider the grader's remarks when they receive them back; their mathematical behavior is shaped, consciously or subconsciously, by these influences. In the next section, I will present a number of examples which I consider symptomatic if not typical, and which might contribute to students' difficulties with explanation and proof. No systematic analysis of textbooks has been carried out; but it is at least conceivable that the given examples are the norm rather than isolated cases.

## 4. Textbooks and Classroom Teaching

The examples which follow should by no means be seen as a critique of the experiences students are subjected to, but simply as a description. They illustrate introductory college and university courses, including service courses. They are likely to be inappropriate for advanced mathematics courses and for transition courses which have been instituted in some places specifically to help beginning mathematics majors make the transition to formal argument (Hillel and Alvarez, 1996).

In many textbooks used at the level under consideration, more or less formal arguments are used, together with visual or intuitive justifications, generic examples, and naive induction. Even the formal arguments are often only formal in appearance. But more importantly, students are rarely if ever given any indications whether mathematics distinguishes between these forms of argumentation or whether they are all equally acceptable.

For example, what does a textbook author expect when he asks students in one exercise to 'show that', and in the next one to 'show by example that' (Anton, 1994). What could and should the student conclude about the expectations author and teacher have in tasks such as Example 6 of Section 2? Indeed, a considerable number of students answered that problem by writing down a specific 2 by 3 matrix for A, a specific 3 by 2 matrix for B, computing AB and BA and, possibly, adding: 'You see!'

A very thoughtfully written recent linear algebra textbook is the one by Lay (1994). It stresses connections between the various concepts and methods usually taught in elementary linear algebra courses wherever possible. But it does not help the student distinguish different forms of justificative argument. For example, after computing the determinant of a 5 by 5 matrix with only two non-zero elements below the diagonal, the author states 'The matrix in Example 3 was nearly triangular. The method in that example is easily adapted to prove the following theorem. THEOREM 2. If A is a triangular matrix, then det A is the product of the entries in the main diagonal of A.' (p. 165) Isn't the student invited to conclude that a computational method, carried out for specific examples, counts as a proof?

Generic examples are used liberally in textbooks; but how often is their role clearly identified, and how often is the range of their genericity and thus their validity discussed?

Experimental and visual arguments are, of course, common in calculus textbooks. An example is Fraleigh (1990) who intuitively introduces the slope of the tangent as limit of the slope of the secant, complete with physical interpretations and with a computer program to compute the derivative, before discussing the notion of limit. The treatment includes statements such as 'The smaller the value of $\Delta x$ (of course $\Delta x = 0$ is not allowed), the better you would expect $m_{sec}$ in Eq. (1) to approximate $m_{tan}$.' (p. 30). This claim is visually supported by a graph with two secants for which the statement is true; it is true in the example at hand but false in general! More precisely, it is false, in general, that the smaller the value of $\Delta x$, the better $m_{sec}$ approximates $m_{tan}$. The qualification 'you would expect' is astutely placed. How would we as teachers react to a student explanation, similar to the ones in Section 2, in which a wrong statement is accompanied by an astutely placed 'you would expect'?

In the next section of Fraleigh's book, the $\epsilon$-$\delta$-definition of limit is introduced intuitively. Later in the book, students are expected to produce proofs, for example to 'Show that a sequence can't converge to two different limits' (Exercise 19, in Section 10.1) and to give $\epsilon$-N-proofs for the convergence of sequences. The transition between the intuitive and the formal stages is not clearly marked. Can and should students be expected to establish the distinction on their own?

And obviously, there is the ubiquitous 'It is easy to see ...' which is missing from few sources – textbooks and research papers alike; how could it fail to lead to explanations such as the following where AY 'sees' that two vectors span $R^3$.

*Example.*
Find the matrix A of the linear transformation T, and determine whether T is onto and/or 1-1-valued:
T: $R^2 \rightarrow R^3$, with

$$T(e_1) = \begin{bmatrix} 3 \\ -6 \\ 0 \end{bmatrix} \qquad T(e_2) = \begin{bmatrix} 1 \\ 2 \\ 5 \end{bmatrix}$$

AY: [correctly constructs T and then continues:] One can see that T spans $R^3$ and therefore the transformation is 'onto'.

In the above cases, choices have presumably been made intentionally and on the basis of didactic considerations by the authors. In other cases, choices may be less obvious and the reasons for them less conscious, even to authors and teachers. For example, many theorems usually taught in calculus courses including the mean value theorem, are based on the completeness of the real numbers. This is usually neither explicitly assumed in calculus textbooks, nor even discussed but taken as an intuitive fait acquis. Many calculus students hardly distinguish between rational and real numbers. Intuitively, the rationals are as complete to them as the reals (Bronner, 1997).

The previous paragraph concerns axiomatics, and one may make the point that mathematicians are often not explicit about their use of axioms, even in research papers. A similar point cannot be made, however, about the next issue, circularity of argumentation. Learning, even in mathematics, often proceeds in an order quite different from the logical one. We learn by establishing connections and relationships, by building a web of ideas rather than a linear and logical sequence of implications; ideas grow synergetically rather than strictly on top of each other. Thus many dilemmas about precedence arise for teachers and textbook authors, for example the dilemma whether to introduce limits before or after derivatives in calculus; the tensions between experimental and rigorous reasoning pointed out above for one calculus text, are common precisely because they are, at least partly, an effect of this dilemma.

This and similar dilemmas lead to circular reasoning on a global level which is usually not easy to identify. Circular reasoning in teaching occurs, however, also at the level of detailed and seemingly rigorous argument, such as the derivation of the important result that $\lim_{x\to 0} \frac{\sin x}{x} = 1$. As pointed out by Richman (1993), this result is equivalent to the inequality sin x < x < tan x (in an appropriate interval) and this inequality is usually established on the basis of an argument which not only uses visual information in a crucial way but also uses the fact that the area of the unit circle is $\pi$; and the area of the unit circle, in turn, is usually established using the exact same inequality that sin x < x < tan x (unless it is taken as given on the authority of the elementary school teacher who happened to teach the students.) Do we, as teachers, have a case in criticizing students like AM (Example 5 in Section 2) for failing to well order his reasons?

Visual, intuitive, generic, experimental, and even circular justifications appear to be common in textbooks. It seems safe to assume that reasoning presented in class is usually less formal than that presented in textbooks. This is certainly so for most classrooms I have visited, including my own. I will illustrate this only by one sequence of events which repeatedly occurred in my own classes before I realized that it is historically documented and may well have occurred similarly in the majority of calculus classrooms taught all over the world during the past 100 years.

Almost every time I taught applications of integration to compute volumes and surfaces, one or two students came up with a variant of the following question. When computing the volume of a rotationally symmetric solid, one builds the corresponding integral after slicing the solid by planes perpendicular to the axis of rotation and approximating the volume of each slice by means of the volume of a 'straight' cylindrical slice of appropriate radius; why is it that the same method gives a wrong result when computing the surface area of such a solid or the length of an arc. (Slanted, conical slices have to be used to obtain the correct surface area, and slanted line segments to obtain the correct arc length.)

The answer I gave was a somewhat vague argument about the remainder (the volume being cut off and neglected) converging to zero in the three-dimensional case but not in the two-dimensional case. Whether the students were satisfied with this answer or not, I don't know – they did not return. Whether I was satisfied, however, I do know. I was not – but I was never able to find a better answer which would still be accessible to the students. And only much later did I realize that I was in very illustrious company. A remark by Young (1969, p. 152) drew my attention to Lebesgue's (1963) booklet 'En marge du calcul des variations' ['At the margin of the Calculus of Variations'] which was probably written in the 1920s. In it, Lebesgue recounts how strongly he has been influenced by an argument which purports to show that the length of one side of the triangle ABC is equal to the sum of the lengths of the two other sides; the argument proceeds by constructing a sequence of broken lines, each of length AB+AC, which approaches BC. Lebesgue writes: 'Tous mes travaux se rattachent à une plaisanterie de collégiens. Au Collège de Beauvais, nous démontrions que, dans un triangle, un côté est égal à la somme des deux autres. ... Mes camarades ne voyaient là qu'une bonne plaisanterie. Pour moi, ce raisonnement m'a paru extrêmement troublant, car je ne voyais aucune différence entre lui et les démonstrations relatives aux aires et surfaces des cylindres, cônes, sphères ou à la longueur de la circonférence' (p. 308). ['All my work is connected to a school boys' joke. At the Collège de Beauvais we proved that in a triangle, one side is equal to the sum of the two others. ... My friends considered this simply a joke. But to me, the argument was extremely troublesome because I did not see any difference between it and the proofs about areas and surfaces of cylinders, cones, spheres or the length of the circumference.'] The high school question under what conditions the length of a curve is the limit of the lengths of infinitely close curves has thus played a central role in Lebesgue's mathematical career.

Most of our students are not precisely like Lebesgue but rather like his classmates. They lack his ability to ask why a mathematical argument is or is not valid. And explanations like the one I gave to those of my students who came up with an excellent question, do little to help them acquire this ability, quite the contrary. I need to ask myself what the status of my explanation for the students is. Why did they accept it? And what does this imply for their willingness to continue to be critical with respect to the mathematical arguments I present? And finally, what right do I have not to accept an argument which is similar in style but given by one of them as answer to my assignment or examination question?

## 5. Theoretical Approaches

Up to this point, we have avoided the theoretical question what constitutes an explanation, and what is a satisfactory, or an acceptable explanation.

We have also avoided a clear distinction between explanation, argument and proof.

There are reasons for avoiding such distinctions, and for stressing commonalties rather than differences. Hanna (1995) has made the point that 'While in mathematical practice the main function of proof is justification and verification, its main function in mathematics education is surely that of explanation' (p. 47). Others require proof to go beyond explanation. For example, when Ellen (Moore, 1994) was asked in a test to 'Prove that if A and B are sets satisfying $A \cap B = A$, then $A \cup B = B$', she wrote: '$A \cap B = A$ says – by definition of intersection – that the members of A and the members of B that are the same are all the members of A. Therefore by definition of subset $A \subseteq B$. If A is a subset of B, all of its members are contained in B. When there is a union of a set and it's subset the union then includes the whole set. Therefore $A \cup B = B$' (p. 258). Moore's interpretation is that 'In contrast to the professor's expectation, Ellen's proof was based on her intuitive understanding ... she did not use the language and rules of inference that had been agreed upon in class. ... Ellen needed to go beyond merely giving an explanation ...' (p. 259).

Moore thus raises the question of the relationship between explanation, proof and understanding. This same question led Sierpinska (1994) to analyze the epistemological differences between explanation and proof in the light of their role in the process of understanding mathematics. Her analysis, based on work of the philosopher Ajdukiewicz, acknowledges '... a close relationship between proving and explaining. Both when proving a theorem and when explaining a state of things we answer to one and the same 'why?' question' (p. 74). She identifies, however, a few important differences. The first is that 'proof aims at increasing the degree of firmness with which we accept a fact as basis for our understanding' (p. 75) whereas explanation 'does not serve as basis for our more positive acceptance of the derived statement'. The second difference is that explanations use examples, models, visualizations and similar means in order to express something about mathematics; explanatory discourse is more metamathematical than mathematical; it may, for example, include reasons why a certain fact is significant in mathematics, something which is clearly beyond the realm of a proof. In this sense, explanation goes beyond proof. Similarly, a proof may call for an explanation which highlights the central idea of the proof. Proof and explanation are thus interwoven in processes of understanding.

Duval (1992–93) takes a similar approach, in that he also uses epistemological and cognitive analysis. He distinguishes three forms of justification: explanation, argument and proof. Two criteria determine the acceptability of the reasons given to justify a claim: their pertinence (semantic coverage between reasons and claim) and their strength (resistance to objections). According to Duval, the main function of an explanation is descriptive; its aim is to produce reasons. Arguments and proofs, on the other hand, examine the strength of these reasons, in particular whether they are free of contradictions; their function is to determine and possibly change the truth value of the claim.

Whereas Sierpinska groups argument and proof together, Duval distinguishes them by their pertinence: In arguments, the semantic content of the reasons is important and determines the epistemic value of the claim; proofs, on the other hand, are detached from content; they must be valid rather than pertinent; the status of a proof, rather than its content, determines its epistemic value. As a consequence, the language of arguments is more natural than that of proofs; the discourse of arguments is globally and thematically coherent. Nevertheless, Duval states that there are no criteria that definitely distinguish arguments from either explanations or proofs. At best, one can use characteristics such as coherence of discourse.

In summary, for mathematics educators there appears to be a continuum reaching from explanation via argument and justification to proof, and the distinctions between the categories are not sharp. Maybe surprisingly, questions as to what constitutes a proof can be and occasionally are asked in mathematics itself. In a review of the history of proof, Kleiner (1991) identified two major themes, namely that the validity of a proof is a reflection of the overall mathematical climate of the time, and that transitions in both directions, from less to more or from more to less rigor usually had good mathematical reasons. Even contemporaries did not always agree on what does and what does not constitute a proof; in particular, formalists and intuitionists were irreconcilably divided over the legitimacy of non-constructive existence proofs at the beginning of the 20th century.

Even more fundamentally, Ernest (1999) notes that 'there is growing recognition that proofs do not follow the explicit rules of mathematical logic, and that acceptance is instead a fundamentally social act'. This recognition is based on recent work by philosophers of mathematics including Lakatos (1978) and Kitcher (1984).

Lakatos (1978) has analyzed the development of analysis at the beginning of the eighteenth century, and in particular Cauchy's proof that the limit of a converging sequence of continuous functions is continuous. How could Cauchy prove this, and publish and maintain his proof, in spite of the fact that he was very well aware of the fact that the limit of a Fourier series can be a step function? It is interesting to learn that according to Cauchy, the sequence does not converge at the jumps; according to Fourier, the limit (step) function is continuous; and according to Abel, Cauchy's 'theorem' had exceptions. Lakatos showed that it took almost 30 years until Cauchy's contemporaries sorted things out sufficiently to see that the confusion arose because their conceptions of the underlying notions, limit and continuity, were not yet sufficiently developed.

Cauchy's argument is convincing; it used the following convergence criterion for series: 'It is sufficient for convergence that, for infinitely large values of the number n, the partial sums S(n), S(n+1), ... differ from the limit S and consequently among themselves by an infinitely small quantity'. This formulation not only uses infinitesimals and infinities in a manner that today makes us raise eyebrows but it is also unclear about the logical status of the crucial phrase 'among themselves'. Example 6 in Section 2 uses a very similar criterion and is equally convincing; it has the additional advantage that the result is true whereas Cauchy's is not, at least not from today's point of view.

In view of these and other recent challenges to the role and status of proof in mathematics (Hanna, 1995; Fallis, 1996; Velleman, 1997) one may legitimately raise the question what, then, do mathematicians consider to be a proof? In an imaginary dialogue between a mathematics professor – the Ideal Mathematician – and a philosophy student who came to ask him what a proof is, Davis and Hersh (1981) convey that the mathematician might recognize a proof when (s)he sees one but is unable to define it or even to improve its description beyond 'Well, it's an argument that convinces someone who knows the subject' (p. 40). The problems inherent in this description are well underlined by the fact the (inadequate) arguments in Examples 7 and 8 (Section 2) did, at least for some time, convince the experts!

Two tasks arise from the theoretical analysis: As didacticians, we must sharpen our awareness of the distinctions between explanation, argument and proof, and we must reflect on what we can and what we should expect from students in different age groups, levels and courses. And as teachers, we must attempt the difficult task of helping students to understand what we expect from them. The examples in Section 2 provide ample room for questioning what is expected by the different formulations used, including 'explain' (Examples 1, 2), 'justify' (Example 3), 'prove' (Example 5), and 'show that' (Example 6). Does 'show that' mean 'formally prove' or 'use an example to demonstrate that' (or something intermediate between these two)? Does 'explain' mean explain to a fellow student or explain in such a way as to convince the teacher that you understand the reasoning behind the claim?

## 6. Students' explanations: What (not) to expect

The two preceding papers in this special issue on forms of knowledge implicitly contribute their share to our understanding why a large part of students' knowledge is not of the kind which supports mathematical justifications: According to Ernest, much of our students' mathematical knowledge is tacit; and while tacit knowledge is likely to be used correctly in applications, it cannot be used explicitly in reasoning. Mason and Spence, on the other hand, show that even students' explicit mathematical knowledge is, to a large extent, not deductive but inductive, abductive or generalized from experience.

As shown in Section 4, teachers and textbooks make extensive use of a great variety of forms of knowledge, and for good reasons. The opportunity to acquire knowledge in a variety of forms, and to establish connections between different forms of knowledge are apt to contribute to the flexibility of students' thinking (Dreyfus and Eisenberg, 1996). The same variety, however, also tends to blur students' appreciation of the difference in status which different means of establishing mathematical knowledge bestow upon that knowledge.

It thus appears that, at least in some measure, the task of learning and teaching mathematical justification conflicts with the pursuit of learning and teaching mathematical relationships, concepts and procedures in a flexible manner. Kline, from whose book (Kline, 1973) I have adapted the title of this paper, has argued convincingly that making logic the guiding principle of curriculum design, as tried by the New Math movement, does not solve this (nor any other) problem. And while recognizing proof as the hallmark of mathematics, he has strongly argued that its place is at the end rather than at the beginning, and that even in proof, rigor should play a lesser role than motivation: 'In no case should one start with the deductive approach, even after students have come to know what this means. The deductive proof is the final step. ... [The student] should be allowed to accept and use any facts that are so obvious to him that he does not realize he is using them. ... Proofs of whatever nature should be invoked only where the students think they are required. The proof is meaningful when it answers the student's doubts, when it proves what is not obvious.' (p. 195).

So where does this leave the students? They have few if any means to distinguish between different forms of reasoning and to appreciate the consequences for the resulting knowledge; nor can they be expected to distinguish between explanation, argument and proof (Section 5). And what means do they have to judge the validity of mathematical arguments? Even for mathematicians it is not always clear cut what a proof is, both philosophically and practically; we should therefore not be surprised if students find it difficult to make such judgments even at the level of the simple, short proofs likely to appear in high school and college classrooms. Hence, there is little reason to be surprised at the findings presented in Section 3 which rather clearly show that most students have at best a very vague notion of what constitutes a mathematical proof.

In spite of this, many teachers, including this author, frequently request students to explain their reasoning, show why a statement is true, justify a claim or even prove a result. What can we realistically expect from students when we ask them to 'explain why', when we ask them to construct an argument? What criteria do we have to judge their productions, and what criteria can we use with good conscience? Is it even realistic to expect high school and college teachers to make judgments and decisions as to whether or not their students' mathematical arguments are acceptable, and to make such judgments in real time in a classroom?

Teachers have to decide how to relate to experimentally based reasoning and to visually based reasoning, and have to adapt their reaction to whether such reasonings are presented as justifications, as explanations or as basis for conjectures. And under what circumstances should one accept a student's justification based on: 'Because my teacher said so'; or 'I can just see it'? My personal experience reported in Section 4 may well lead to a 'because the teacher said so'; and we have all experienced students 'seeing' blatantly wrong things. But what if the conclusion which the student 'can see' is correct? What about a student who 'can see' that 26/65 equals 2/5? Maybe the student 'can see' the digit 6 disappear from numerator and denominator? So correctness of the answer is not the issue, certainly not the main issue.

On the other hand teachers are often willing to accept the slightest sign of a student's understanding as satisfactory explanation, even if the student's words and actions leave much to be desired. The teacher may recognize that the student has more or less consciously established some connections between what is given and what is to be justified. In other cases, visual reasoning can be deep and go far beyond vaguely seeing some connections. The term 'visual reasoning' is used here to refer to arguments based on analysis of a diagrammatic situation (Dreyfus, 1994). Visual reasoning is often analytic in the sense that the thinking subject consciously analyzes the visual images, and reflects on them. Such reasoning may include analyzing, acting on and transforming images, mental or external ones, and drawing conclusions about mathematical relationships from these actions. It is apt to underlie detailed justifications of mathematical statements, even rigorous proofs (Barwise and Etchemendy, 1995).

The question under what conditions, and according to which criteria, visually based explanations can and should be accepted has received little attention and it is thus left up to the individual teacher to intuitively decide on a case to case basis.

The situation is similar with respect to experimentally based reasoning – and the reference is not to the use of experiment in exploration, and in the generation of conjectures but in students' justificative explanations. Should a teacher accept the argument that a sequence converges because numerical experiment shows so? Or rather, in which situations should the teacher accept such an argument? And if not, what arguments are acceptable? Is a Cauchy type argument such as in Example 4 of Section 2 preferable? Why? And when? Should students be allowed to use infinitesimals in a similar manner as Cauchy did? Why or why not? How about a visual argument showing how successive elements of the sequence come closer to each other? How close does such a visual argument need to mimic a proof in the Weierstrass (epsilon-delta) sense? And to what extent, in what respect, does an epsilon-delta argument reveal more (or less) about a student's understanding than an experimental one?

Just as for visual arguments, the question arises under what conditions experimentally based explanations can and should be accepted. What criteria can and should a teacher apply and what general considerations can help mathematics educators and teachers to establish such criteria?

In conclusion, the requirement to explain and justify their reasoning requires students to make the difficult transition from a computational view of mathematics to a view that conceives of mathematics as a field of intricately related structures. This implies acquiring new attitudes and conceiving of new tasks: The central question changes from 'What is the result?' to 'Is it true that . . .?'. Students thus need to develop new and more sophisticated forms of knowledge.

Although it has been known for some time how complex and difficult this transition is, only a few attempts to directly deal with it have been reported in the literature (e.g., Movshowitz-Hadar, 1988; Dreyfus and Hadas, 1996), and even these have made little or no attempt to assess changes in students' views of mathematics and their ability to explain and justify. The question how to sensitize students to this change and help them achieve it, remains open.

Of equal importance, and equally open is the development of criteria which can be used by teachers to judge the acceptability of their students' mathematical arguments, and of principles on which the development and examination of such criteria can be based.

## REFERENCES


Alibert, D. and Thomas, M.: 1991, 'Research on mathematical proof', in D. Tall (ed.), *Advanced Mathematical Thinking,* Kluwer, Dordrecht, The Netherlands, pp. 215–230.

Anton, H.: 1994, *Elementary Linear Algebra,* Wiley, New York, USA.

Arnold, V. I.: 1991, 'A Mathematical Trivium', *Russian Mathematical Surveys* 46(1), 271–278.

Balacheff, N.: 1987, 'Processus de preuve et situations de validation', *Educational Studies in Mathematics* 18(2), 147–176.

Barwise, J. and Etchemendy, J.: 1995, *Hyperproof,* CSLI Lecture Notes, No. 42, Stanford University, Stanford, CA, USA.

Bronner, A.: 1997, 'Les rapports d'enseignants de troisième et de seconde aux objets 'nombre réel' et 'racine carrée', *Recherches en didactique des mathématiques* 17(3), 55–80.

Cauchy, A.-L.: 1821, *Cours d'analyse de l'école royale polytechnique*, de Bure, Paris, France.

Coe, R. and Ruthven, K.: 1994, 'Proof practices and constructs of advanced mathematics students', *British Educational Research Journal* 20(1), 41–53.

Davis, P. J. and Hersh, R.: 1981, *The Mathematical Experience,* Birkhäuser, Boston.

Dreyfus, T.: 1994, 'Imagery and reasoning in mathematics and mathematics education', in D. Robitaille, D. Wheeler and C. Kieran (eds.), *Selected Lectures from the 7th International Congress on Mathematical Education,* Les presses de l'université Laval, Sainte-Foy, Québec, Canada, pp. 107–122.

Dreyfus, T.: in press, 'La demostracion a lo largo del curriculum' ('Experiencing proof throughout the curriculum'), in J. Deulofeu and N. Gorgorio (eds.), *Educación matemática: retos y cambios desde una perspectiva internacional*, ICE/GRAO pub., Barcelona, Spain.

Dreyfus, T. and Eisenberg, T.: 1996, 'On different facets of mathematical thinking', in R. J. Sternberg and T. Ben-Zeev (eds.), *The Nature of Mathematical Thinking*, Lawrence Erlbaum, Mahwah, NJ, USA.

Dreyfus, T. and Hadas, N.: 1996, 'Proof as answer to the question why', *Zentralblatt für Didaktik der Mathematik* 28(1), 1–5.

Duval, R.: 1992–1993, 'Argumenter, démontrer, expliquer: continuité ou rupture cognitive?', *petit x* 31, 37–61.

Ernest, P.: 1999, 'Forms of knowledge in mathematics and mathematics education: Philosophical and rhetorical perspectives', *Educational Studies in Mathematics* 38, 67–83.

Fallis, D.: 1996, 'Mathematical Proof and the Reliability of DNA Evidence', *American Mathematical Monthly* 103(6), 491–497.

Ferrari, P. L.: 1997, 'Action-based strategies in advanced algebraic problem solving', in E. Pehkonen (ed.), *Proceedings of the Twenty-First International Conference on the Psychology of Mathematics Education,* University of Helsinki, Finland, Vol. 2, pp. 257–264.

Finlow-Bates, K., Lerman, S. and Morgan, C.: 1993, 'A survey of current concepts of proof help by first year mathematics students', in I. Hirabayashi, N. Nohda, K. Shigematsu and F.-L. Lin (eds.), *Proceedings of the Seventeenth International Conference on the Psychology of Mathematics Education,* University of Tsukuba, Japan, Vol. I, pp. 252–259.

Fischbein, E.: 1982, 'Intuition and proof', *For the Learning of Mathematics* 3(2), 9–24.

Fraleigh, J. B.: 1990, *Calculus with Analytic Geometry,* Addison-Wesley, Menlo Park, CA, USA.

Hanna, G.: 1995, 'Challenges to the importance of proof', *For the Learning of Mathematics* 15(3), 42–49.

Harel, G. and Sowder, L.: 1998, 'Students' proof schemes: Results from Exploratory Studies', in E. Dubinsky, A. H. Schoenfeld and J. J. Kaput (eds.), *Research on Collegiate Mathematics Education*, Vol. III, American Mathematical Society, Providence, RI, USA, pp. 234–283.

Hillel, J. and Alvarez, J. C.: 1996, 'University mathematics, report of topic group 3′, in C. Alsina, J. Alvarez, M. Niss, A. Perez, L. Rico and A. Sfard (eds.), *Proceedings of the 8th International Congress on Mathematical Education*, S.A.E.M. 'Thales', Sevilla, pp. 245–248.

Kitcher, P.: 1984, *The Nature of Mathematical Knowledge,* Oxford University Press, New York, USA.

Kleiner, I.: 1991, 'Rigor and proof in mathematics: a historical perspective', *Mathematics Magazine* 64(5), 291–314.

Kline, M.: 1973, *Why Johnny Can't Add,* Random House, New York, USA.

Lakatos, I.: 1978, 'Cauchy and the continuum', *The Mathematical Intelligencer* 1(3), 151–161.

Lampert, M.: 1990, 'When the problem is not the question and the solution is not the answer: mathematical knowing and teaching', *American Educational Research Journal* 27(1), 29–63.

Lay, D. C.: 1994, *Linear Algebra and Its Applications*, Addison-Wesley, Reading, MA, USA.

Lebesgue, H.: 1963, En marge du calcul des variations', *L'enseignement mathématique* 9(4), 212–326.

Maher, C. A. and Martino, A. M.: 1996, 'The development of the idea of mathematical proof: a 5-year case study', *Journal for Research in Mathematics Education* 27(2), 194–214.

Margolinas, C.: 1992, 'Eléments pour l'analyse du rôle du maître: les phases de conclusion', *Recherches en Didactique des Mathématiques* 12(1), 113–158.

Martin, G. and Harel, G.: 1989, 'Proof frames of preservice elementary teachers', *Journal for Research in Mathematics Education* 20(1), 41–51.

Mason, J. and Spence, M.: 1999, 'Beyond mere knowledge of mathematics: The importance of Knowing-to Act in the moment', *Education Studies in Mathematics* 38, 135–161.

Moore, R. C.: 1994, 'Making the transition to formal proof', *Educational Studies in Mathematics* 27(3), 249–266.

Movshowitz-Hadar, N.: 1988, 'Stimulating presentation of theorems followed by responsive proofs', *For the Learning of Mathematics* 8(2), 12–19.

Richman, F.: 1993, 'A Circular Argument', *The College Mathematics Journal* 24(2), 160–162.

Sierpinska, A.: 1994, *Understanding in Mathematics,* Falmer Press, London, UK.

Silver, E. A.: 1994, 'Mathematical thinking and reasoning for all students: moving from rhetoric to reality', in D. Robitaille, D. Wheeler and C. Kieran (eds.), *Selected Lectures from the 7th International Congress on Mathematical Education,* Les presses de l'université Laval, Sainte-Foy, Québec, Canada, pp. 311–326.

Velleman, D. J.: 1997, 'Fermat's last theorem and Hilbert's program', *The Mathematical Intelligencer* 19(1), 64–67.

Vinner, S.: 1997, 'The pseudo-conceptual and the pseudo-analytical thought processes in mathematics learning', *Educational Studies in Mathematics* 34(2), 97–129.

Yackel, E. and Cobb, P.: 1996, 'Sociomathematical norms, argumentation, and autonomy in mathematics', *Journal for Research in Mathematics Education* 27(4), 458–477.

Young, L. C.: 1969, *Calculus of Variations and Optimal Control Theory*, Saunders, Philadelphia, USA.

Zack, V.: 1997, ' "You have to prove us wrong": Proof at the elementary school level', in E. Pehkonen (ed.), *Proceedings of the Twenty-First International Conference on the Psychology of Mathematics Education,* University of Helsinki, Finland, Vol. 4, pp. 291–298.

*Center for Technological Education, Holon*
*(affiliated with Tel Aviv University)*
*P.O. Box 305, Holon 58102, Israel*
*E-mail: nttommy@weizmann.weizmann.ac.il*

EDDIE GRAY, MARCIA PINTO, DEMETRA PITTA and DAVID TALL

# KNOWLEDGE CONSTRUCTION AND DIVERGING THINKING IN ELEMENTARY & ADVANCED MATHEMATICS

ABSTRACT. This paper begins by considering the cognitive mechanisms available to individuals which enable them to operate successfully in different parts of the mathematics curriculum. We base our theoretical development on fundamental cognitive activities, namely, *perception* of the world, *action* upon it and *reflection* on both perception and action. We see an emphasis on one or more of these activities leading not only to different kinds of mathematics, but also to a spectrum of success and failure depending on the nature of the focus in the individual activity. For instance, geometry builds from the fundamental *perception* of figures and their shape, supported by action and reflection to move from practical measurement to theoretical deduction and euclidean proof. Arithmetic, on the other hand, initially focuses on the *action* of counting and later changes focus to the use of symbols for both the process of counting and the concept of number. The evidence that we draw together from a number of studies on children's arithmetic shows a divergence in performance. The less successful seem to focus more on perceptions of their physical activities than on the flexible use of symbol as process and concept appropriate for a conceptual development in arithmetic and algebra. Advanced mathematical thinking introduces a new feature in which concept definitions are formulated and formal concepts are constructed by deduction. We show how students cope with the transition to advanced mathematical thinking in different ways leading once more to a diverging spectrum of success.

## 1. Constructing Mathematical Knowledge

Mathematical development occurs in a biological brain. To enable a structure with complex simultaneous activity to pursue sequential thought in a coherent way requires a special mechanism. Crick suggests:

> The basic idea is that early processing is largely parallel: a lot of different activities proceed simultaneously. Then there appear to be one or more stages where there is a bottleneck in information processing. Only one (or a few) 'object(s)' can be dealt with at a time. This is done by temporarily filtering out the information coming from the unattended objects. The attention system then moves fairly rapidly to the next object, and so on, so that attention is largely serial (i.e. attending to one object after another) not highly parallel (as it would be if the system attended to many things at once). (Crick, 1994, p. 61)

The powerful thinking that develops in mathematics takes advantage of this biological phenomenon. The filtering out of most activity to focus on a few elements requires that these elements be distilled to their essence so that

*Educational Studies in Mathematics* **38:** 111–133, 1999.

they are 'small enough' to be considered at one time. It also requires that each of these elements be appropriately linked to other relevant structures in the huge memory store to allow it speedily to become a new focus of attention as required.

One method to cope with the complexity of a sequence of activities is repetition and practice until it becomes routine and can be performed with little conscious thought. This frees the conscious memory to focus on other items (Skemp, 1979). For instance, in using tools, the techniques become part of unconscious activity whilst the individual can focus on more utilitarian or aesthetic issues. Although such repetition and interiorisation of procedures has been seen as an essential part of mathematics learning, for decades it has been known that it has made no improvement in the understanding of relationships (see for example, Thorndike, 1922; Brownell, 1935). More importantly, if used exclusively, it may lead to a form of procedural thinking that lacks the flexibility necessary to solve novel problems (see for example, Schoenfeld, 1992).

A more powerful method of dealing with complexity is through the human use of language. Here a single word can stand not only for a highly complex structure of concepts and/or processes but also for various levels in a conceptual hierarchy. Perception of figures is at the foundation of geometry, but it takes the power of language to make hierarchical classifications. Figures are initially perceived as gestalts but then may be described and classified through verbalising their properties, to give the notions of points, lines, planes, triangles, squares, rectangles, circles, spheres, etc. Initially these words may operate at a single generic level, so that a square (with four *equal* sides and every angle a right angle) is not considered as a rectangle (with only *opposite* sides equal). Again, through verbal discussion, instruction and construction, the child may begin to see hierarchies with one idea classified within another, so that 'a square is a rectangle is a quadrilateral', or 'a square is a rhombus is a parallelogram is a quadrilateral'. The physical and mental pictures supported by linguistic descriptions may become conceived in a more pure, imaginative way. Points have 'position but no size', straight lines are truly straight, with 'no thickness and arbitrary length', a circle is the locus of a point a fixed distance from the centre and so on. Such a development leads to platonic *mental* constructions of objects and the development of Euclidean geometry and Euclidean proof. Thus, a focus on perceived objects leads naturally through the use of language to platonic mental images and a form of mathematical proof (as in Van Hiele, 1959, 1986).

On the other hand, the idea of counting begins with the repetition of number words, with the child's remembered list of numbers steadily growing in length and correctness of sequence. The act of counting involves pointing at successive objects in a collection in turn and saying the number words, 'there are one, two, *three* things here.' This may be compressed, for instance, by carrying out the count silently, saying just the last word, 'there are [one, two,] three', heard as 'there are ... three.' It is thus natural to use the word 'three' not just as a counting word, but also as a number concept. By this simple device, the counting process 'there are one, two, three,' is compressed into the concept 'there are three.' (Gray and Tall, 1994).

This compression is powerful in quite a different way from the compression in geometric thinking. In geometry, a word represents a generic concept (say 'square') in a hierarchy of concepts. In arithmetic the number word is also part of a hierarchy (a counting number is a fraction is a rational number is a real number). However, the major biological advantage of numbers arises not from this hierarchy but from the way in which the number words can be used to switch between *processes* (such as counting or measuring) and *concepts* (such as numbers). Not only are number symbols 'small enough' to be held in the focus of attention as concepts, they also give immediate access to action schemas (such as counting) to carry out appropriate computations. In the biological design of the brain, they act not only as economical units to hold in the focus of attention, they also provide direct links to action schemas.

When numbers have become conceived as mental entities, they may themselves be operated upon. For instance, two numbers may be added to give their sum through a development that again involves a process of compression. The addition of two numbers begins as 'count-all', involving three counting stages: 'count one set, count another, put them together and count them all'. This is compressed through various stages including 'count-on', where the first number is taken as the starting value and the second is used to count-on to give the result. Some of these results are committed to memory to give 'known facts'. They may then be used in a conceptual way to 'derive facts', for instance, knowing that $5 + 5$ is 10, to deduce that $5 + 4$ is one less, namely, 9.

This power of mathematical symbols to evoke either process or concept caused Gray and Tall (1994) to give the notion a formal name. The amalgam of a *process*, a *concept* output by that process, and a *symbol* that can evoke either process or concept is called a *procept*. In elementary arithmetic, procepts start as simple structures and grow in interiority with the cognitive growth of the child. Although other theorists (including Dubinsky, 1991; and Sfard, 1991) use the term 'object', we prefer the word 'concept' because terms such as 'number concept' or 'fraction concept' are more common in ordinary language than 'number object' or 'fraction object'. Further, the term is used in a manner related to the 'concept image' which consists of 'all of the mental pictures and associated properties and processes' related to the concept in the mind of the individual (Tall and Vinner, 1981, p. 152). Procepts are generic and increase in richness with the growing sophistication of the learner. There is no claim that there is a 'thing' called 'a mental object' in the mind. Instead, a symbol is used which can be *spoken, heard, written* and *seen*. It has the distilled essence that can be held in the mind as a single entity, it can act as a link to internal action schemas to carry out computations, and it can be communicated to others.

### 1.1. *Piaget's three forms of abstraction*

Piaget spoke of three forms of abstraction. When acting on objects in the external world, he speaks first of *empirical abstraction*, where the focus is on the *objects* themselves and 'derives its knowledge from the properties of objects' (Beth and Piaget, 1966, pp. 188–189). On the other hand, a focus on the *actions* leads to *pseudo-empirical abstraction* which 'teases out properties that the action of the subjects have introduced into objects' (Piaget, 1985, pp. 18–19). Further constructions can then be accomplished by *reflective abstraction*, using existing structures to construct new ones by observing one's thoughts and abstracting from them. In this way:

> ... the whole of mathematics may therefore be thought of in terms of the construction of structures, ... mathematical entities move from one level to another; an operation on such 'entities' becomes in its turn an object of the theory, and this process is repeated until we reach structures that are alternately structuring or being structured by 'stronger' structures. (Piaget, 1972, p. 703)

Note here that reflective abstraction seems to be formulated as a mental version of 'pseudo-empirical abstraction', in which an 'operation' on a mental entity becomes in its turn an 'object' at the next level. Some authors (for example, Dubinsky, 1991) have taken this to mean that reflective abstraction only occurs by processes becoming conceived as conceptual entities through a process of 'encapsulation' or 'reification'. Given Piaget's two notions of abstraction from the physical world, the question naturally arises as to whether there are corresponding forms of reflective abstraction focusing on mental objects and on mental actions. Our analysis would support this position. In the cognitive development of geometry, there is a clear shift from the mental conception of a physical triangle to the mental construction of a perfect platonic triangle. The former is imagined drawn on paper, with lines having thickness joining points having size, the latter has perfectly straight edges with no thickness and vertices with position but no size. We therefore suggest that there are (at least) *two* forms of reflective abstraction, one focusing on *objects*, occurring, for instance, in Euclidean geometry, the other focusing on *actions on objects* (usually represented by symbols), for instance, in arithmetic, algebra and the calculus.

Our focus on perception, action and reflection is therefore consistent with Piaget's three notions of abstraction, with the additional observation that reflective abstraction has a form which focuses on objects and their properties, as well as one which focuses on actions and their encapsulation as objects.

### 1.2. *Theories of process-object transformation*

The notion of (dynamic) processes becoming conceived as (static) objects has played a central role in various theories of concept development (see, for example, Dienes, 1960; Piaget, 1972; Greeno, 1983; Davis, 1984; Dubinsky, 1991; Sfard, 1991; Harel and Kaput, 1991; Gray and Tall, 1994).

Dubinsky and his colleagues (e.g. Cottrill et al., 1996) formulate a theory which they give the acronym APOS, in which *actions* are physical or mental transformations on objects. When these actions become intentional, they are characterised as *processes* that may be later encapsulated to form a new *object*. A coherent collection of these actions, processes and objects, is identified as a *schema*. In more sophisticated contexts, empirical evidence also intimates that a schema may be reflected upon and acted on, resulting in the schema becoming a new object through the encapsulation of cognitive processes (Cottrill et al., 1996, p. 172).

Sfard (1991, p. 10) suggests that 'in order to speak about mathematical *objects*, we must be able to deal with the products of some *processes* without bothering about the processes themselves'. Thus we begin with 'a process performed on familiar objects' (Sfard and Linchevski, 1994, p. 64). This is then 'condensed' by being seen purely in terms of 'input/output without necessarily considering its component steps' and then 'reified' by converting 'the already condensed process into an object-like entity.' Sfard postulates her notion of 'reification' within a wider theory of *operational* and *structural* conceptions, the first focusing on processes, the second on objects (Sfard, 1989, 1991; Sfard and Linchevski, 1994). In several papers she emphasises that the operational approach – constructing new objects through carrying out processes on known objects usually precedes a structural approach to the new objects themselves.

Such theories, which see the construction of new mental objects through actions on familiar objects, have a potential flaw. If objects can only be constructed from cognitive actions on already established objects, where do the initial objects come from?

Piaget's theory solves this problem by having the child's preliminary activities involving perception and action of the physical world. Once the child has taken initial steps in empirical or pseudo-empirical abstraction to construct mental entities, then these become available to act upon to give a theoretical hierarchy of mental constructions.

Sfard's theory concentrates on later developments in older individuals who will already have constructed a variety of cognitive objects. Dubinsky also concentrates on undergraduate mathematicians. However, the APOS theory is formulated to apply to all forms of object formation. Dubinsky, Elterman and Gong (1988, p. 45), suggest that a 'permanent object' is constructed through 'encapsulating the process of performing transformations in space which do not destroy the physical object'. This theory therefore follows Piaget by starting from initial *physical* objects that are not part of the child's cognitive structure and theorises about the construction of a *cognitive* object in the mind of the child. It formulates empirical abstraction as another form of process-object encapsulation.

At the undergraduate level, Dubinsky (1991) extends APOS theory to include the construction of axiomatic theories from formal definitions. APOS theory is therefore designed to formulate a theory of encapsulation covering all possible cases of mental construction of cognitive objects.

Our analysis has different emphases. We see the differences between various types of mathematical concept formation being as least as striking as the similarities. For instance, the construction of number concepts (beginning with pseudo-empirical abstraction) follows a very different cognitive development from that of geometric concepts (beginning with empirical abstraction) (Tall, 1995). In elementary mathematics, we see two different kinds of cognitive development. One is the van Hiele development of geometric objects and their properties from physical perceptions to platonic geometric objects. The other is the development of symbols as process and concept in arithmetic, algebra and symbolic calculus. It begins with actions on objects in the physical world, and requires the focus of attention to shift from the action of counting to the manipulation of number symbols. From here the number symbols take on a life of their own as cognitive concepts, moving on to the extension and generalisations into more sophisticated symbol manipulation in algebra and calculus. Each shift to a new conceptual domain involves its own subtle changes and cognitive reconstructions, however, what characterises these areas of elementary mathematics is the use of symbols as concepts and processes to calculate and to manipulate.

### 1.3. *A new focus in advanced mathematical thinking*

When formal proof is introduced in advanced mathematical thinking, a new focus of attention and cognitive activity occurs. Instead of a focus on symbols and computation to give answers, the emphasis changes to selecting certain properties as definitions and axioms and building up the other properties of the defined concepts by logical deduction. The student is often presented with a context where a formal concept (such as a mathematical group) is encountered both by examples and by a definition. Each of the examples satisfies the definition, but each has additional qualities, which may, or may not, be shared between individual examples. The properties of the formal concept are deduced as theorems, thus constructing meaning for an overall umbrella concept from the concept definition. This didactic reversal – constructing a mental object from 'known' properties, instead of constructing properties from 'known' objects causes new kinds of cognitive difficulty.

The new formal context – in which objects are created from properties (axioms) instead of properties deduced from (manipulating) objects – not only distinguishes advanced mathematical thinking from elementary mathematical thinking, it also suggests that different kinds of 'structure' occur in the structural-operational formulation of Sfard. In elementary mathematics, for example, a 'graph' is described as a structural object (Sfard, 1991). In advanced mathematics, the Peano postulates are said to be structural (Sfard, 1989). Thus, a structural perspective may refer to visual objects in elementary mathematics and Bourbaki-style formal structure in advanced mathematics.

### 1.4. *A theoretical perspective*

The preceding discussion leads to a theory of cognitive development in mathematics with two fundamental focuses of attention – *object* and *action* – together with the internal process of *reflection*. In line with Piaget we note the different forms of *abstraction* which arise from these three: *empirical abstraction*, *pseudo-empirical abstraction* and *reflective abstraction*. However, we note that reflective abstraction itself has aspects that focus on object or on action.

We see abstraction from physical objects as being different from abstraction from actions on objects. In the latter case, action-process-concept development is aided by the use of symbol as a pivot linking the symbol either to process or to concept. Procepts occur throughout arithmetic, algebra and calculus, and continue to appear in advanced mathematical thinking. However, the introduction of axioms and proofs leads to a new kind of cognitive concept – one which is *defined* by a concept definition and its properties *deduced* from the definition. We regard the development of formal concepts as being better formulated in terms of the definition-concept construction. This focuses not only on the complexity of the definition, often with multiple quantifiers, but also on the internal conflict between a concept image, which 'has' properties, and a formal concept, whose properties must be 'proved' from the definitions.

We therefore see elementary mathematics having two distinct methods of development, one focusing on the properties of *objects* leading to geometry, the other on the properties of *processes* represented symbolically as procepts. Advanced mathematics takes the notion of *property* as fundamental, using properties in concept definitions from which a systematic formal theory is constructed.

## 2. Diverging Cognitive Development in Elementary Mathematics

### 2.1. *Divergence in performance*

The observation that some individuals are more successful than others in mathematics has been evident for generations. Piaget provided a novel method of interpreting empirical evidence by hypothesising that all individuals pass through the same cognitive stages but at different paces. Such a foundation underlies the English National Curriculum with its sequence of levels through which all children pass at an appropriate pace, some progressing further than others during the period of compulsory education.

Krutetskii (1976, p. 178) offers a different conception with a spectrum of performance between various individuals depending on how they process information. He studied 192 children selected by their teachers as 'very capable' (or 'mathematically gifted'), 'capable', 'average' and 'incapable'. He found that gifted children remembered general strategies rather than detail, curtailed their solutions to focus on essentials, and were able to provide alternative solutions. Average children remembered specific detail, shortened their solutions only after practice involving several of the same type, and generally offered only a single solution to a problem. Incapable children remembered only incidental, often irrelevant detail, had lengthy solutions, often with errors, repetitions and redundancies, and were unable to begin to think of alternatives.

Our research also shows a divergence in performance. We do not use the evidence collected to imply that some children are doomed forever to erroneous procedural methods whilst others are guaranteed to blossom into a rich mathematical conceptualisation. We consider it vital not to place an artificial ceiling on the ultimate performance of any individual, or to predict that some who have greater success today will continue to have greater success tomorrow. However, the evidence we have suggests that the different ways in which individuals process information at a given time can be either beneficial or severely compromising for their current and future development. A child with a fragmented knowledge structure who lacks powerful compressed referents to link to efficient action schemas will be more likely to have greater difficulty in relating ideas. The expert may see distilled concepts which can each be grasped and connected within the focus of attention. The learner may have diffuse knowledge of these conceptual structures which is not sufficiently compressed into a form that can be brought into the focus of attention at a single time for consideration.

Far from not working hard enough, the unsuccessful learner may be working very hard indeed but focusing on less powerful strategies that try to cope with too much uncompressed information. The only strategy that may help them is to rote-learn procedures to perform as sequential action schemas. Such knowledge can be used to solve routine problems requiring that particular technique, but it occurs *in time* and may not be in a form suitable for thinking about as a whole entity.

### 2.2. *Focus on objects and/or actions in elementary mathematics*

The observation that a divergence in performance exists in the success and failure of various students does not of itself explain *how* that divergence occurs. To gain an initial insight into aspects of this divergence, we return to our initial notions of perception, action and abstraction. We earlier discussed global differences between geometry (based on perception of figures, supported by action and extended through reflection), and arithmetic (based on actions of counting objects that are initially perceived and reflected upon). Now, within arithmetic we consider the effect of different emphases on action, perception and reflection. Whenever there is an activity involving actions on objects, the complexity of the activity may cause the individual to focus only on part of the activity. For instance, it is possible to focus on the objects, on the actions or a combination of the two. Cobb, Yackel and Wood (1992) see this attention to objects or actions as one of the great problems in learning mathematics, particularly if learning and teaching are approached from a representational context. Pitta and Gray (1997) showed that certain observed differences in children's arithmetic performance could be linked to the learner's focus either on objects, on actions, or on a combination of both.

To investigate the way in which children may focus on different aspects of a situation, Pitta (1998) placed five red unifix cubes before some seven-year-old children at the extremes of mathematical ability. She asked the children to indicate what they thought about when they saw the cubes and what they thought would be worth remembering about them. The four more able children all had something to say about the cubes using the notion of 'five'. They all thought that 'five cubes' was worth remembering. In contrast, the four lower ability children talked about the pattern, the colour, or the possible rearrangements of the cubes and considered these to be worth remembering.

Different contexts require a focus of attention upon different things. Within an art lesson it may be important to filter out those things that may not immediately be seen to be part of an aesthetic context. Number may be one of these. In the mathematical context it is important to filter out those things that may not be seen to be mathematical. Yet, in the activity just considered, low achievers seemed less able to do this, continuing to focus on their concrete experience. High achievers, on the other hand, were able to separate the inherent mathematical qualities from the actual physical context. They could also, if required, expand their discussions to include other aspects of the activity, revealing cognitive links to a wider array of experience. Such differences may become manifest in the way in the activity is remembered. It is hypothesised that low achievers focus upon the physical aspects of the activity, which are assimilated in an episodic way. High achievers appear to focus upon the semantic mathematical aspects, which are accommodated in a generic way (Pitta and Gray, 1997).

### 2.3. *The proceptual divide*

The divergence in success between extremes of success and failure can be usefully be related to the development of the notion of procept. Gray and Tall (1994) suggest that interpretations of mathematical symbolism as process or procepts leads to a *proceptual divide* between the less successful and the more successful. On the one hand, we see a cognitive style strongly associated with invoking the use of procedures, on the other a style more in tune with the flexible notion of procept. Those using the latter have a cognitive advantage; they derive considerable mathematical flexibility from the cognitive links relating process and concept. In practice, there is a broad spectrum of performance between different individuals in different contexts (Figure 1).

In a given routine context, a specific procedure may be used for a specific purpose. This allows the individual to *do* mathematics in a limited way, provided that it involves using the learned procedure. Some individuals may develop greater sophistication by being able to use alternative procedures for the same process and to select a more efficient proced-

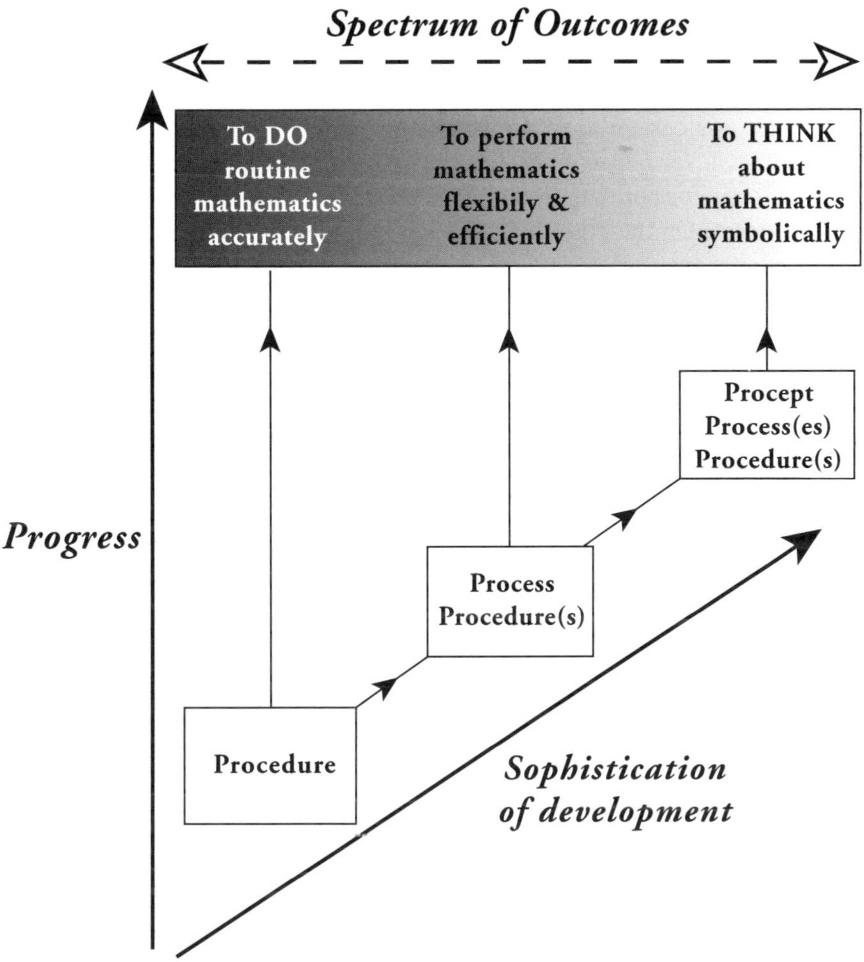


*Figure 1.* A spectrum of performance in using mathematical procedures, processes, procepts

ure to carry out the given task speedily and accurately. For instance, the procedure of 'count-on from largest' is a quicker way of solving $2 + 7$ (counting on 2 after 7 rather than counting-on 7 after 2). Baroody and Ginsburg (1986) suggest that growing sophistication arises from the recognition that a single mathematical process may be associated with several procedures. Woods, Resnick and Groen (1975) note that this element of 'choice' can be indicative of increased sophistication. However, it is only when the symbols used to represent the process are seen to represent manipulable concepts that the individual has the proceptual flexibility both to *do* mathematics and also to mentally manipulate the concepts at a more sophisticated level (Gray and Tall, 1994).

In a particular case, all three levels (procedure, process, procept) might be used to solve a given routine problem. It might therefore be possible for individuals at different levels of sophistication to answer certain questions in a test at a certain level. However, this may be no indication of success at a later level because the procept in its distilled manipulable form is more ready for building into more sophisticated theories than step-by-step procedures. On the other hand, all too frequently, children are seen using procedures even when they are inappropriate, inefficient and unsuccessful (see for example, Gray, 1993). Those who operate successfully at the procedural level are faced with much greater complexity than their proceptual colleagues when the next level of difficulty is encountered.

### 2.4. *Mental representations and elementary mathematics*

The notion of a proceptual divide illustrates the extreme outcomes of different cognitive styles. We now turn to asking *why* such a difference occurs. To gain a partial answer to this question we now consider mental representations, particularly those in imaginistic form.

Pitta and Gray (1997) describe the way in which two groups of children, 'low achievers' and 'high achievers', report their mental representations when solving elementary number combinations. Differences that emerged showed the tendency of low achievers to concretise numbers and focus on detail. Their mental representations were strongly associated with the procedural aspects of numerical processing – action was the dominant level of operating (see also Steffe, Von Glasersfeld, Richards and Cobb, 1983). In contrast, high achievers appeared to focus on those abstractions that enable them to make choices.

The general impression was that children of different levels of arithmetical achievement were using qualitatively different objects to support their mathematical thinking. Low achievers translated symbols into numerical processes supported by the use of imaginistic objects that possessed shape and in many instances colour. Frequently they reported mental representations strongly associated with the notion of number track although the common object that formed the basis of each 'unit' of the track was derived from fingers. In some instances children reported seeing full picture images of fingers, in others it was 'finger like'. The essential thing is that the object of thought was 'finger' and the mental use of finger invoked a double counting procedure. The objects of thought of the low achievers were analogues of perceptual items that seemed to force them to carry out procedures in the mind, almost as if they were carrying out the procedures with perceptual items on the desk in front of them. Pitta and Gray suggest that their mental representations were essential to the action; and they maintained the focus of attention. When items became more difficult, the children reverted to the use of real items.

In contrast, when high achievers indicated that they had 'seen something', that 'something' was usually a numerical symbol. More frequently these children either responded automatically or reported that they talked things over in their heads. However, when they did describe mental representations the word 'flashing' often dominated their description. Representations came and went very quickly. 'I saw '3 + 4' flash through my mind and I told you the answer', 'I saw a flash of answer and told you.' It was not unusual for the children to note that they saw both question and answer 'in a flash', sometimes the numerical symbol denoting the answer 'rising out of' the symbols representing the question. In instances where children reported the use of derived facts it was frequently the numerical transformation that 'flashed'. For instance when given 9 + 7 one eleven year old produced the answer 16 accompanied by the statement. '10 and 6 flashed through my mind'. Here we have vivid evidence of powerful mental connections moving from one focus of attention to another. Such a child evidently has flexible mental links between distilled concepts that allow quick and efficient solutions to arithmetic problems.

This ability to encapsulate arithmetical processes as numerical concepts provides the source of flexibility that becomes available through the proceptual nature of numerical symbolism. Recognising that a considerable amount of information is compressed into a simple representation, the symbol, is a source of mathematical power. This strength derives from two abilities; first an ability to filter out information and operate with the symbol as an object and secondly the ability to connect with an action schema to perform any required computation. We suggest that qualitative differences in the way in which children handle elementary arithmetic may be associated with their relative success. Different cognitive styles seem to indicate that differing perceptions of tasks encountered lead to different consequences, one associated with *performing* mathematical computations, the other associated with *knowing* mathematical concepts.

Mental representations associated with the former appear to be products of reflection upon the actions and the objects of the physical environment. One consequence of mathematical activity focusing upon procedural activity is that it would seem to place a tremendous strain on working memory. It does not offer support to the limited space available within short-term memory.

## 3. The Transition to Advanced Mathematical Thinking

> The move from elementary to advanced mathematical thinking involves a significant transition: that from *describing* to *defining*, from *convincing* to *proving* in a logical manner based on those definitions. ... It is the transition from the *coherence* of elementary mathematics to the *consequence* of advanced mathematics, based on abstract entities which the individual must construct through deductions from formal definitions. (Tall, 1991, p. 20)

The cognitive study of 'advanced mathematical thinking' developed in the mathematics education community in the mid-eighties (see for example, Tall, 1991). Euclidean proof and the beginnings of calculus are usually considered 'advanced' at school level. However, the term 'advanced mathematical thinking' has come to focus more on the thinking of creative professional mathematicians imagining, conjecturing and proving theorems. It is also applied to the thinking of students presented with the axioms and definitions created by others. The cognitive activities involved can differ greatly from one individual to another, including those who build from images and intuitions in the manner of a Poincaré and those more logically oriented to symbolic deduction such as Hermite.

Piaget's notion of 'formal operations' indicates the ability to reason in a logical manner:

> Formal thought reaches its fruition during adolescence ... from the age of 11–12 years ... when the subject becomes capable of reasoning in a hypothetico-deductive manner, i.e., on the basis of simple assumptions which have no necessary relation to reality or to the subject's beliefs, and ... when he relies on the necessary validity of an inference, as opposed to agreement of the conclusions with experience. (Piaget, 1950, p. 148)

In a similar manner, the SOLO taxonomy identifies the formal mode of thinking where:

> The elements are abstract concepts and propositions, and the operational aspect is concerned with determining the actual and deduced relationships between them; neither the elements nor the operations need a real-world referent. (Collis and Romberg, 1991, p. 90)

However, often these ideas are applied by Piaget to imagined *real-world* events and in the SOLO taxonomy to logical arguments in traditional algebra, involving relationships between symbols that no longer need have a perceptual referent.

The notion of advanced mathematical thinking is more subtle than this. It involves the creation of new mental worlds in the mind of the thinker which may be entirely hypothetical. Mathematicians do this by reflecting on their visual and symbolic intuitions to suggest useful situations to study, then to specify *criteria* that are necessary for the required situation to hold. This is done by formulating *definitions* for mathematical concepts as a list of axioms for a given structure, then developing other properties of this structure by deduction from the definitions. A considerable part of research effort is expended in getting these criteria precise so that they give rise to the required deduced properties. What is then produced is more than a verbal/symbolic list of definitions and theorems. Each individual theoretician develops a personal world of concept images and relationships related to the theory. These may include ideas that suggest what *ought* to be true in the given theory before necessarily being able to formulate a proof of what *must* follow from the definitions.

Definitions of structures – such as 'group', 'vector space', 'topological space', 'infinite cardinal' – face in two ways. They face *back* to previous experiences which suggest what ideas are worth studying and *forward* to the construction of theorems which are true for any structure that satisfies the given criteria. They can cause great cognitive problems for a learner who must distinguish between those things in the mind which *suggest* theorems and other things that have already been *proved* from the criteria. The learner must maintain a distinction between the broad concept images formed from previous experience and new constructions – the *formal concept image* – which consists only of those concepts and properties that have been constructed formally from the definitions.

In practice, this often proves extremely difficult. Whereas mathematics researchers may have had experience at *making* new structures by constructing their own definitions, students are more likely to only be initially involved in *using* definitions which have been provided by others. Through their earlier life experiences they will have developed an image in which objects are 'described' in words in terms of collecting together enough information to identify the object in question for another individual. The idea of giving a verbal definition as a list of criteria and *constructing* the concept from the definition is a reversal of most of the development in elementary mathematics where mathematical objects are thought to *have* properties which can be discovered by studying the objects and related processes. The move from the *object* → *definition* construction to *definition* → *object* construction is considered an essential part of the transition from elementary to advanced mathematical thinking.

This definition → object construction involves selecting and using criteria for the definitions of objects. This may reverse previous experiences of relationships. For instance, the child may learn of subtraction as an operation before meeting negative numbers and inverse operations. In formal mathematics the axioms for an additive operation in a group may specify the inverse $-a$ of an element $a$ and define subtraction $b - a$ as the sum of $b$ and $-a$. In this way the presentation of axiom systems as criteria for theoretical mathematical systems can strike foreign chords in the cognitive structure of the learner. Instead of proving results of which they are unsure by starting from something they know, they find they are trying to prove something they know starting from axioms which make them feel insecure.

Our experience of this learning process in mathematical analysis (Pinto and Gray, 1995; Pinto and Tall, 1996; Pinto, 1996, 1998) shows a spectrum of student performances signalling success and failure through following two complementary approaches.

One approach, which we term 'natural' (following Duffin and Simpson, 1993) involves the student attempting to build solely from his or her own perspective, attempting to give meaning to the mathematics from current cognitive structure. Successful natural learners can build powerful formal structures supported by a variety of visual, kinaesthetic and other imagery, as in the case of student Chris (Pinto, 1998). He made sense of the definition of convergence by drawing a picture and interpreting it as a sequence of actions:

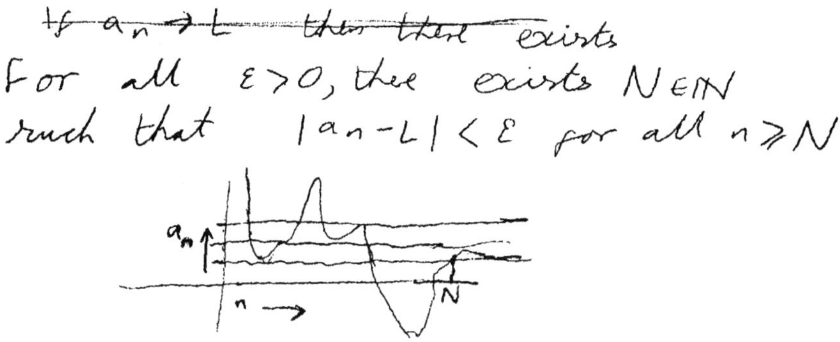


I think of it graphically ... you got a graph there and the function there, and I think that it's got the limit there ... and then $\epsilon$, once like that, and you can draw along and then all the ... points after $N$ are inside of those bounds. ... When I first thought of this, it was hard to understand, so I thought of it like that's the $n$ going across there and that's $a_n$. ... Err, this shouldn't really be a graph, it should be points. (Chris, first interview)

As he drew the picture, he gestured with his hands to show that first he imagined how close he required the values to be (either side of the limit), then how far he would need to go along to get all successive values of the sequence inside the required range. He also explained:

I don't memorise that [the definition of limit]. I think of this [picture] every time I work it out, and then you just get used to it. I can nearly write that straight down. (Chris, first interview)

However, his building of the concept involved him in a constant state of reconstruction as he refined his notion of convergence, allowing it to be increasing, decreasing, up and down by varying amounts, or constant in whole or part, always linking to the definition which gave a single unifying image to the notion. During his reconstructions, he toyed with the idea of an increase in $N$ causing a resultant reduction in the size of $\epsilon$, before settling on the preference for specifying $\epsilon$, then finding an appropriate $N$.

As an alternative to the 'natural' approach, there is a second approach which Pinto (1998) termed 'formal'. Here the student concentrates on the definition, using it and repeating it as necessary until it can be written down without effort. Ross, for example explained he learned the definition:

> Just memorising it, well it's mostly that we have written it down quite a few times in lectures and then whenever I do a question I try to write down the definition and just by writing it down over and over again it gets imprinted and then I remember it. (Ross, first interview)

He wrote:

A sequence $(a_n)$ tends to limit $L$ if, $\forall\, \varepsilon>0,\ \exists\, N(\varepsilon) \in \mathbb{N}$

s.t. $\forall\, n \geq N$;

$$|a_n - L| < \varepsilon.$$

The focus in this case is on the definition and the deductions. Visual and other images play a less prominent role. Used successfully, this approach can produce a formal concept image capable of using the definitions and proving theorems as required by the course. At its very best the student will also be in a position later on to reconstruct knowledge, comparing old with new and making new links. However, it is also possible to develop the knowledge in a new compartment, not linked to old knowledge.

Both formal and natural learners can be successful in advanced mathematical thinking. However, they face different sequences of cognitive reconstruction. The natural learner may be in continuous conflict as (s)he reconstructs informal imagery to give rich meaning to the formal theory. The formal learner may have fewer intuitions to guide the way, but follows a course involving more new construction rather than reconstruction. At the end of the formalisation process, if the new knowledge is linked to the old imagery, then reconstruction is likely to be required at this stage.

Less successful students also have difficulties in different ways. Some (such as those in Pinto and Gray, 1995) saw the new ideas only in terms of their old meanings and could not make the transition to the use of definition as criteria for determining the concept. These could be described as natural learners who fail to reconstruct their imagery to build the formalism. Their informal concept image intimates to them that the theorems are 'true' and they see no need to support informal imagery with what they regard as alien to both their need and their understanding.

Less successful students attempting the formal route may be unable to grasp the definition as a whole and cope with only parts of it. They may be confused by the complexity of multiple quantifiers, perhaps failing to give them their true formal meaning, perhaps confusing their purpose, perhaps concentrating only on a part of the definition.

It seems that the only way out for unsuccessful students, be they natural or formal learners, is to attempt to rote-learn the definitions.

> Maths education at university level, as it stands, is based like many subjects on the system of lectures. The huge quantities of work covered by each course, in such a short space of time, make it extremely difficult to take it in and understand. The pressure of time seems to take away the essence of mathematics and does not create any true understanding of the subject. From personal experience I know that most courses do not have any lasting impression and are usually forgotten directly after the examination. This is surely not an ideal situation, where a maths student can learn and pass and do well, but not have an understanding of his or her subject. *Third Year Mathematics Student*, (Tall, 1993a)

## 4. Conclusion

In this paper, we have considered the interplay of perception, action and reflection on cognitive development in mathematics. Geometry involves a major focus on perception of *objects*, which develops through reflective activity to the mental construction of perfect platonic objects. Arithmetic begins by focusing on *actions* on objects (counting) and develops using procepts (symbols acting as a pivot between processes and concepts) to build elementary arithmetic and algebra.

In elementary arithmetic we find that the less successful tend to remain longer focused on the nature of the objects, their layout and the procedures of counting. Our evidence suggests that less successful children focus on the specific and associate it to real and imagined experiences that often do not have generalisable, manipulable aspects. We theorise that this places greater strains on their overloaded short-term memory. A focus on the counting procedure itself can give limited success through procedural methods to solve simple problems. High achievers focus increasingly on flexible proceptual aspects of the symbolism allowing them to concentrate on mentally manipulable concepts that give greater conceptual power. The flexible link between mental concepts to think about and action schemas to do calculations utilise the facilities of the human brain to great advantage.

We see the transition to advanced mathematical thinking involving a transposition of knowledge structure. Elementary mathematical concepts *have* properties that can be determined by acting upon them. Advanced mathematical concepts are *given* properties as axiomatic definitions and the nature of the concept itself is built by deducing the properties by logical deduction. Students handle the use of concept definitions in various ways. Some natural learners reconstruct their understanding to lead to the formal theory whilst other, formal, learners build a separate understanding of the formalities by deduction from the concept definitions. However, many more can make little sense of the ideas, either as natural learners whose intuitions make the formalism seem entirely alien, or as formal learners who cannot cope with the complexity of the quantified definitions.

The theory we present here has serious implications in the teaching of elementary and advanced mathematics, in ways which have yet to be widely tested. The obvious question to ask is 'how can we help students acquire more beneficial ways of processing information?', in essence, 'how can we help those using less successful methods of processing to become more successful?' Our instincts suggest that we should attempt to *teach* them more successful ways of thinking about mathematics. However, this strategy needs to be very carefully considered, for it may have the result that we teach procedural children flexible thinking *in a procedural way*. This scenario would have the effect of burdening the less successful child with even more procedures to cope with. It might tend to make their cognitive structure *more complex* rather than *more flexible and more efficient*.

One approach at encouraging more flexible thinking (Pitta and Gray, 1997a,b) used a graphic calculator with a multi-line display retaining several successive calculations for a child to use in a learning experiment. The experience was found to have a beneficial effect in changing the mental imagery of a child who previously experienced severe conceptual difficulty. Before using the calculator, the child's arithmetic focused on counting using perceptual objects or their mental analogues. After a period of approximately six months use with the graphic calculator, it was becoming clear in our interactions with her that she was associating a different range of meanings with numbers and numerical symbolism. She was beginning to build new images, symbolic ones that could stand on their own to provide options that gave her greater flexibility. The evidence suggests that if practical activities focus on the process of evaluation and the meaning of the symbolism they may offer a way into arithmetic that helps those children who are experiencing difficulty.

In the teaching of algebra, Tall and Thomas (1991) found that the act of programming could allow students to give more coherent meaning to symbolism as both process and concept. A computer language will evaluate expressions, so that, for instance, the learner may explore the idea that $2 + 3 * x$ usually gives a different answer from $(2 + 3) * x$ for numerical values of $x$. This can provide a context for discussing the ways in which expressions are evaluated by the computer. The fact that $2*(x+3)$, $2*x+2*3$, $2*x+6$, always give the same output, can be explored to see how different procedures of evaluation may lead to the same underlying process, giving the notion of equivalent expressions and laying down an experiential basis for manipulating expressions. This leads through a procedure – process – procept sequence in which expressions are first procedures of evaluation, then processes which can have different expressions producing the same effect, then concepts which can themselves be manipulated by replacing one equivalent expression by another.

In advanced mathematical thinking more research is required to test whether different methods of approach may support different personal ways to construct (and reconstruct) formal theory. Just as Skemp (1976) referred to the difficulty faced by a relational learner taught by instrumental methods (or vice versa), we hypothesises that there are analogous difficulties with natural learners being taught by formal methods (or vice versa). This suggests that more than one approach is required to teaching mathematical analysis. Some students may benefit from a study quite different from the traditional formal theory. For example, Tall (1993b) observed that a class of student teachers similar to those who failed to make any sense of the formalism (see Pinto and Gray, 1995) could construct natural insights into highly sophisticated ideas using computer visualisations even though this may not improve their ability to cope with the formal theory.

Success can be achieved for some students in various ways. These include giving meaning to the definitions by reconstructing previous experience, or by extracting meaning from the definition through using it, perhaps memorising it, and then building meaning within the deductive activity itself (Pinto 1998). However, not all succeed. Those who fail are often reduced, at best, to learning theorems by rote to pass examinations. How different this is from the advanced mathematical thinking of the creative mathematician, with its combination of intuition, visualisation and formalism combined in different proportions in different individuals to create powerful new worlds of mathematical theory.

## REFERENCES


Baroody, A.J. and Ginsburg, H.P.: 1986, 'The relationship between initial meaningful and mechanical knowledge of arithmetic', in J. Hiebert (ed.), *Conceptual and Procedural Knowledge: The Case for Mathematics*, Lawrence Erlbaum Associates, Hillsdale, NJ, pp. 75–112.

Beth, E.W. and Piaget, J.: 1966, *Mathematical Epistemology and Psychology*, (translated by W. Mays), D. Reidel. Dordrecht, The Netherlands (originally published 1965).

Brownell, W.A.: 1935, 'Psychological considerations in the learning and teaching of arithmetic', in W.D. Reeve (ed.), *Teaching of Arithmetic, The Tenth Yearbook of the National Council of Teacher's of Mathematics*, Bureau of Publication, Teachers College, Columbia University.

Cobb, P., Yackel, E. and Wood, T.: 1992, 'A constructivist alternative to the representational view of mind in mathematics education', *Journal for Research in Mathematics Education* 23, 2–23.

Collis., K. and Romberg, T.: 1991, 'Assessment of mathematical performance: An analysis of open-ended test items', in C. Wittrock and E. Baker (eds.), *Testing and Cognition*, Prentice-Hall, New Jersey, pp. 82–116.

Cottrill, J., Dubinsky, E., Nichols, D., Schwingendorf, K., Thomas, K. and Vidakovic, D.: 1996, 'Understanding the limit concept: Beginning with a co-ordinated process schema', *Journal of Mathematical Behaviour* 15, 167–192.

Crick, F.: 1994, *The Astonishing Hypothesis*, Simon & Schuster, London.

Davis, R.B.: 1984, *Learning mathematics: the cognitive science approach to mathematics education*, Ablex. Publishing Co., Norwood, NJ.

Dienes, Z.P.: 1960, *Building up Mathematics*, Hutchinson Educational, London.

Dubinsky, E.: 1991, 'Reflective abstraction', in D.O. Tall (ed.), *Advanced Mathematical Thinking*, Kluwer Academic Publishers, Dordrecht, The Netherlands, pp. 95–123.

Dubinsky, E., Elterman, F. and Gong, C.: 1988, 'The students construction of quantification', *For the Learning of Mathematics* 8, 44–51.

Duffin, J.M. and Simpson. A.P.: 1993, 'Natural, conflicting, and alien', *Journal of Mathematical Behaviour* 12, 313–328.

Gray, E.M.: 1993, 'Count-on: The parting of the ways in simple arithmetic', in I. Hirabayashi, N. Hohda, K. Shigematsu and Fou-Lai Lin (eds.), *Proceedings of XVII International Conference on the Psychology of Mathematics Education*, Tsukuba, Japan, Vol. I, pp. 204–211.

Gray, E.M. and Tall, D.O.: 1994, 'Duality, ambiguity and flexibility: A proceptual view of simple arithmetic', *Journal for Research in Mathematics Education* 25(2), 115–141.

Gray, E.M. and Pitta, D.: 1997, 'Changing Emily's Images', *Mathematics Teaching* 161, 38–51.

Greeno, J.: 1983, 'Conceptual entities', in D. Genter and A.L. Stevens (eds.), *Mental Models*, Lawrence Erlbaum, Hillsdale, NJ, pp. 227–252.

Harel, G. and Kaput, J.: 1992, 'Conceptual entitities in advanced mathematical thinking: The role of notations in their formation and use', in D.O. Tall (ed.), *Advanced Mathematical Thinking*, Kluwer Academic Publishers, Dordrecht, The Netherlands, pp. 82–94.

Krutetskii, V.A.: 1976, *The Psychology of Mathematical Abilities in Schoolchildren*, (translated by J. Teller), University of Chicago, Chicago.

Piaget, J.: 1950, *The Psychology of Intelligence*, (translated by M. Piercy), Routledge and Kegan Paul, London.

Piaget, J.: 1972, *The Principles of Genetic Epistemology*, (translated by W. Mays), Routledge & Kegan Paul, London.

Piaget, J.: 1985, *The Equilibrium of Cognitive Structures*, Harvard University Press, Cambridge Massechusetts.

Piaget, J. and Inhelder, B.: 1971, *Mental Imagery in the Child*, Basic, New York.

Pinto, M.M.F.: 1996, *Students' Use of Quantifiers*, Paper presented to the Advanced Mathematical Thinking Working Group at The Twentieth Conference of the International Group for the Psychology of Mathematics Education, Valencia, Spain.

Pinto, M.M.F.: 1998, 'Students' Understanding of real analysis', Unpublished Doctoral Thesis, Mathematics Education Research Centre, University of Warwick, UK.

Pinto, M.M.F. and Gray, E.: 1995, 'Difficulties teaching mathematical analysis to non-specialists', in D. Carraher and L. Miera (eds.), *Proceedings of X1X International Conference for the Psychology of Mathematics Education*, Recife, Brazil, 2, 18–25.

Pinto, M.M.F. and Tall, D.O.: 1996, 'Student teachers' conceptions of the rational numbers', in L. Puig and A. Guitiérrez (eds.), *Proceedings of XX International Conference for the Psychology of Mathematics Education*, Valencia, 4, pp. 139–146.

Pitta, D. and Gray, E.: 1997a, 'In the Mind. What can imagery tell us about success and failure in arithmetic?', In G.A. Makrides (ed.), *Proceedings of the First Mediterranean Conference on Mathematics*, Nicosia, Cyprus, pp. 29–41.

Pitta, D. and Gray, E.: 1997b, 'Emily and the supercalculator', in E. Pehkonen (ed.), *Proceedings of XXI International Conference for the Psychology of Mathematics Education*, Lahti, Finland, 4, pp. 17–25.

Pitta, D.: 1998, 'In the mind. Internal representations and elementary arithmetic', Unpublished Doctoral Thesis, Mathematics Education Research Centre, University of Warwick, UK.

Schoenfeld, A.H.: 1992, 'Learning to think mathematically: Problem solving, metacognition, and sense making in mathematics', in D.A. Grouws (ed.), *Handbook of research on mathematics teaching and learning: A project of the National Council of Teachers of Mathematics*, MacMillan, New York, pp. 334–370.

Sfard, A.: 1989, 'Transition from operational to structural conception: The notion of function revisited', in G. Vergnaud, J. Rogalski, M. Arigue (eds.), *Proceedings of XIII International Conference for the Psychology of Mathematics Education,* Paris, France, Vol. 3, pp. 151–158.

Sfard, A.: 1991, 'On the dual nature of mathematical conceptions: Reflections on processes and objects as different sides of the same coin', *Educational Studies in Mathematics* 22, 1–36.

Sfard, A. and Linchevski, L.: 1994, 'The gains and pitfalls of reification–the case of algebra', *Educational Studies in Mathematics* 26, 191–228.

Skemp, R.R.: 1976, 'Relational understanding and instrumental understanding', *Mathematics Teaching* 77, 20–26.

Skemp, R.R.: 1979, *Intelligence, Learning and Action*, Chichester, U.K., John Wiley & Sons.

Steffe, L., Von Glaserfeld, E., Richards, J. and Cobb, P.: 1983, *Children's Counting Types: Philosophy, Theory and Applications*, Preagar, New York.

Tall, D.O.: 1991, *Advanced Mathematical Thinking*, Kluwer Academic Publishers, Dordrecht, The Netherlands.

Tall, D.O.: 1993a 'Mathematicians thinking about students thinking about mathematics', *Newsletter of the London Mathematical Society* 202, 12–13.

Tall, D.O.: 1993b, 'Real mathematics, rational computers and complex people', *Proceedings of the Fifth Annual International Conference on Technology in College Mathematics Teaching*, pp. 243–258.

Tall, D. O.: 1995, 'Cognitive growth in elementary and advanced mathematical thinking', in D. Carraher and L. Miera (eds.), *Proceedings of XIX International Conference for the Psychology of Mathematics Education*, Recifc, Brazil. Vol. 1, pp. 61–75.

Tall, D.O. and Vinner, S.: 1981, 'Concept image and concept definition in mathematics with particular reference to limits and continuity', *Educational Studies in Mathematics* 12, 151–169.

Thorndike, E.L.: 1922, *The Psychology of Arithmetic*, MacMillan, New York.

Van Hiele, P. and D.: 1959, *The child's thought and geometry*, Reprinted (1984), in D. Fuys, D. Geddes and R. Tischler (eds.), *English translation of selected writings of Dina van Hiele-Geldof and Pierre M. van Hiele*, Brooklyn College, Brooklyn, NY, pp. 1–214.

Van Hiele, P.: 1986, *Structure and Insight*, Academic Press, Orlando.

Woods, S.S., Resnick, L.B. and Groen, G.J.: 1975, 'An experimental test of five process models for subtraction', *Journal of Educational Psychology* 67, 17–21.

EDDIE GRAY, MARCIA PINTO, DEMETRA PITTA and DAVID TALL
*Mathematics Education Research Centre,*
*University of Warwick,*
*Coventry CV4 7AL,*
*U.K.*
*E-mail: E.M.Gray@warwick.ac.uk*

JOHN MASON and MARY SPENCE

# BEYOND MERE KNOWLEDGE OF MATHEMATICS: THE IMPORTANCE OF KNOWING-TO ACT IN THE MOMENT

ABSTRACT. *Knowing-to* is active knowledge which is present in the moment when it is required. To try to produce knowing-to, formal education focuses on forms of knowing which are easier to teach and to test: knowing-that (factual), knowing-how (technique and skills), and knowing-why (having a story in order to structure actions and from which to reconstruct actions). Together these constitute knowing-about the subject. Expertise is demonstrated by being able to respond to assessments: to write essays and to solve routine problems. The central problem of education is that knowing-about does not in itself guarantee knowing-to, as teachers have attested throughout the ages. For example, Edward Fitzgerald (Harrison, 1937) captures it beautifully in one stanza of his purported translation of the Rubaiyat of Omar Khayyam:

Myself when young did eagerly frequent,
Doctor and Saint and heard great argument,
About it and about: but ever more
Came out by the same door as in I went (p. 341).

Instead of trying to reach definitions, we illustrate distinctions amongst kinds of *knowing* as used by various authors in the past. Then we turn to our own experience, for it is in one's own experience that one can locate and enliven sources of metaphoric resonances and metonymic triggers which constitute understanding. Drawing on our experience we distinguish knowing-to from other forms of knowing, and explore implications of that distinction for teaching and learning mathematics. We propose that *knowing-to* act in the moment depends on the structure of attention in the moment, depends on what one is aware of. Educating this awareness is most effectively done by labelling experiences in which powers have been exhibited, and developing a rich network of connections and triggers so that actions 'come to mind'. No-one can act if they are unaware of a possibility to act; no-one can act unless they have an act to perform.

## 1. Introduction

'My students don't seem to know what they studied last week, much less last month or last year!'
When marking examinations: 'These students don't seem to have learned much of anything!'

Teachers find it frustrating when they discover that their assumptions about what students know in the way of facts, topics, theorems, and techniques

*Educational Studies in Mathematics* **38:** 135–161, 1999.

studied in previous courses prove to be ill-founded. Often students will deny ever having encountered ideas that the teacher knows they met the previous year. Marking student examinations can be depressing because students rarely seem to do as well as expected. Why are these experiences so constant from generation to generation?

Our answer is that what students know tends to be compartmentalised (for example fractions and decimals are seen as distinct entities) whereas what really matters are connections between topics. Students caught up in knowing facts and techniques do not automatically develop the awareness which would enable them to know-to use that knowledge in new situations.

Students can often write good essays in response to assignments, and in the case of mathematics can often solve routine problems of the type on which they have been trained. However as soon as students are given something more general or less familiar, or a task requiring several steps, they are mostly at sea. They don't appear to know-to use what they have learned. For example, Hugh Burkhardt often suggests in seminars that it is unreasonable to expect students to use tools for mathematical modelling that they first encountered in the previous two or three years, and descriptions of the difficulty students have with multi-step problems are legion. Brown, Collins and Duguid (1989) observed that '... students may pass exams ... but still not be able to use a domain's cognitive tools.', which echoes Whitehead's observation that 'It is possible to acquire a tool but to be unable to use it' (Whitehead, 1932). For Whitehead 'the problem of keeping knowledge alive, of preventing it from becoming inert' (p. 7) was the central problem of all education. It takes time to integrate tools into your own functioning, to have them become 'ert' (as opposed to inert).

Active, practical knowledge, knowledge that enables people to act creatively rather than merely react to stimuli with trained or habituated behaviour involves *knowing-to* act, in the moment. This is what students need in order to engage in problem solving where context is novel and resolution non-routine or multi-layered; this is what teachers need in order to provoke students into educating their awareness as well as training their behaviour. Bereiter and Scardammalia (1989) reach a similar if somewhat implicit conclusion: 'in order to learn what is ostensibly being taught in school, students need to direct mental effort to goals over and above those implicit in school activities'. Although teachers believe they are teaching students to know actively, their experience, as expressed in the opening quotes, suggests otherwise.

## 2. Knowing-About

The English language is dominated by prepositions, or so it seems to non-native speakers. Prepositions provide subtleties of nuance which differ from verb to verb. Prepositions used with the verb *to know*, and, more importantly, with the gerund *knowing*, which implies a continuing process rather than a static state, offer subtle distinctions that are quite hard to communicate and yet, we shall argue, they direct and circumscribe the process of education.

### 2.1. *Historical-cultural perspectives*

Traditionally, philosophers have seen knowledge as 'justified true belief', but in examining more closely what each of these primitive terms means (*justified*, *true*, and *belief*), either an inherent circularity is exposed, or else these primitive terms in turn become problematic. We see knowledge not as an approximation to some fixed or absolute state of justified true belief, but as at best a snapshot of a state of knowing that is in constant flux according to prevailing personal and social conditions.

Classifying types of knowledge has an ancient pedigree: Plato engaged in it in his many Socratic dialogues, and Locke (1894) outlines a variety of *sorts* and *degrees* of knowledge in his monumental essay on human understanding. Russell (1914, pp. 35, 151), following in the footsteps of Hume, distinguished *knowledge by acquaintance*, achieved directly through sense impressions, from *knowledge by description*, which involves expressing acquaintance in language. For example, a social practice reconstructed by an individual from being immersed in its use by others in a community is embodied knowing, a form of knowledge by acquaintance, as is knowledge gained by 'doing things oneself'. By contrast, institutionalised knowledge 'conveyed' through formal educational processes (largely verbal) is knowledge by description. Mathematics as a practice depends on both insight, and expression of that insight to others. Russell's distinction can be recast as a distinction between *knowledge of* and *knowledge that*, mirroring French usage of *connaître* and *savoir*. Islamic philosophers similarly distinguish *acquired knowledge* from *knowledge by presence,* the latter including immediate and direct grasp of the real, and thus encompassing even more than Russell's *acquaintance* (Honderich, 1995, p. 448).

In his seminal work, Ryle (1949) distinguished between *knowing-that* (factually), *knowing-how* (to perform acts), and *knowing-why* (having stories to account for phenomena and actions). Burton (1995) suggests that knowing-that has an impersonal connotation, but that when knowing-that is extended to knowing-why, that is, to having personal stories to account for, to make connections with, and to reconstruct from, then mathematics becomes personal and more idiosyncratic. Biggs (1994) rehearses Ryle's distinctions and then directs attention to what he considers to be a different way of considering knowledge, in five hierarchical levels following Bruner's spiral curriculum (Bruner, 1966) and resonant of van Hiele's levels (1986):

*Tacit*: manifested through doing without conscious awareness or stories

*Intuitive*: directly perceived or felt

*Declarative*: description of how and why expressed in some symbol system that is publicly understandable

*Theoretical*: abstracted or generalised statements going beyond particular instances

*Metatheoretical*: knowledge about the process of abstraction and theory building

Following a pragmatic, even pragmaticist line, De Jong and Ferguson-Hessler (1996) take knowledge-in-use as their orientation for a comprehensive analysis of the plethora of constructs through which people have made distinctions between different types of knowledge. They use categories such as *situational knowledge*, *conceptual knowledge*, *procedural knowledge*, and *strategic knowledge*, and qualities such as *level* (as in surface or deep), *connectedness, automatedness, modality* (as in verbal and pictorial), and *generality* (as against domain specific) to generate a matrix of types against qualities. Despite different words and different parts of speech, all these categories have much in common with Ryle's basic distinctions.

Skemp (1979, p. 259) distinguished *knowing-that* and *knowing-how* from *knowing-to*, which he described in terms of *being able to* use a technique in a fresh situation. However, his use of the term *ability* shows how difficult it is to remain with the core experiential-phenomenological nature of *knowing-to*, and how easy it is to get caught in more absolute sounding terms (either one has the ability or one does not). The term *ability* and its currently popular cognate *competency* carry implications of ontological commitment to something objective and invariant, whereas knowing-to is for us, by its very nature, fleeting, unstable, and context-situation dependent.

Knowing-about, that is, knowing-that, -how, and -why forms the heart of institutionalised education: *students* can learn and be tested on it. But success in examinations gives little indication of whether that knowledge can be used or called upon when required, which is the essence of knowing-to. Although knowing-to does of course depend on training in behaviour, it is based, as we shall see, in awareness. It has to do with the structure of attention.

## 2.2. *Approaches to knowing from a teacher's perspective*

Knowing has been approached from a teacher's perspective by many authors, using a variety of terms in more or less technical sense. Ordinary words used technically rarely retain precise meaning, for each author can find distinctions in what was previously seen holistically. For example, Kieren (1994) quotes a technical report of Miller, Malone and Kandl (1992) as using Ryle's terms to cover a spectrum of meanings

*knowing-that*, ranging from discrete to integrated knowing;
*knowing-how,* ranging from simple to complex, and
*knowing-why,* ranging from the intuitive to the rigorous.

They used these three types of knowing as structure for their discussion of teacher's perceptions of the meaning of understanding. These three knowings also figure prominently in the different components of teacher knowledge identified by Shulman (1987):

subject matter content knowledge,
pedagogical content knowledge,
knowledge of other related content,
knowledge of curriculum,
knowledge of learners,
knowledge of educational aims,
general pedagogical knowledge.

Shulman's forms of knowledge are supposed to equip the effective practitioner to act, but knowing-to act when the moment comes requires more than having accumulated knowledge-about. It requires relevant knowledge to come to the fore so it can be acted upon. That is what *knowing-to* captures for us. In Mason and Spence (1998) we argue that knowing-to act requires awareness, and that it is working on this awareness which provides the fulcrum for professional development.

Fennema and Franke (1992) review research on teacher's knowledge and its impact on teaching and learning, but again from a static, possessive stance rather than from a dynamic and evolving one. Consequently the issue about whether knowledge is called upon when appropriate is dealt with in terms of deficiencies and inadequacies rather than in terms of generation, evolution, co-emergence and the psychological conditions which produce awareness in the moment when it is needed. Maturana and Varela (1972, 1988) and Maturana (1988), see knowledge as a construct of an observer trying to account both for co-emergent evolution of actions and interactions, and for concensual coordination of actions amongst people, between people and things, and between people and ideas.

## 3. Knowledge and Understanding

Knowledge is often thought to lead to, or be concomitant with, understanding. In mathematics, Sierpinska (1994) considered understanding variously as an act, as experience, as process and as a way of knowing, and showed that links between acts of teaching and student understanding are far from robust. Pirie and Kieren (1989) develop a complex model of layers of mathematical understanding, with flowing out and folding back as observable movements in a spiralling, almost recursive sequence of revisitations, from increasingly enriched perspectives, of previously met ideas. Although understanding provides an alternative description for the sort of knowledge which enables people to act sensibly and with inspiration in the moment, there is still a difference: you can 'understand' but not know-to act, for example Frayn's analysis of Heisenberg's visit to Bohr in 1941, expanded upon later; you can know-to-act and yet not fully understand, for example children acting as though they are using a theorem, such as commutativity of addition, but who of course are not articulate about it Vergnaud (1981).

### 3.1. *The influence of theories about knowledge and knowing*

The views one has about knowledge reflexively influence what is known, and in particular, knowing or not knowing to act. A view of knowledge as static and accreted as sufficient for knowing to act leads to different study practices from a view of knowing as dynamic, situated, and evolving. In their review of students views about knowledge and learning, Hofer and Pintrich (1997) observe that despite considerable research since Piaget's genetic epistemology (Piaget, 1950)

> '... there is still very little agreement on the actual construct under study, the dimensions it encompasses, whether epistemological beliefs are domain specific or how such beliefs might connect to disciplinary beliefs, and what linkages there might be to other constructs in cognition and motivation.'

In their attempt to find some common ground and directions for further research, Hofer and Pintrich review Perry's notion of epistemological positions (Perry, 1968) and developments from it concerning gender differences, to arrive at justification and reasoning as the key issues. This brings us full circle back to the notion of knowledge as justified true belief.

### 3.2. *Why knowledge becomes or remains inert*

Chevellard's *transposition didactique* (Chevellard, 1985; Kang and Kilpatrick, 1992) together with Brousseau's *contrat didactique* (Brousseau, 1984, 1997) develop Whitehead's inertness of taught knowledge, explaining why it is that institutionalisation of education dwells in knowing-about: what is taught is what can be tested, so expert awareness is transformed into instruction in behaviour (*transposition didactique*); students expect that they will learn simply by 'doing' the tasks the teacher sets, and the more clearly and precisely the teacher specifies the behaviour sought, the easier it is for students to display that behaviour without generating it from their own understanding (*contrat didactique*). Consequently all the forces are towards inert knowledge as trained behaviour. Whitehead (1932) forcefully asserted that

> 'this evil path represented by a book or set of lectures which will practically enable a student to learn by heart all the questions likely to be asked ... culminates in a uniform examination [which] is so deadly' (p. 7–8).

In their review of the literature on inert knowledge as a common phenomenon Renkl, Mandl and Gruber (1996) locate three types of explanation for knowledge being inert: *metaprocess* (disturbance to accessing what is needed), *structure deficit* (aspects missing in what is known), and *situatedness* (mismatch between current situation and previous situations). In the last category they note particularly Clancey (1993), and Greeno, Smith and Moore (1993) who see knowledge as relationally defined and not a property of the individual. Greeno et al. use the analogy of motion: motion is not a property of an object, because it depends upon frame of reference; motion is a relation between a frame of reference and an object. So too, knowing is not a property of a person, but of a person in a situation.

In our view, situatedness is all too easily over-stressed at the expense of individual psychology. Although as we shall argue, the situation as experienced triggers and resonates access to acts, those triggers are psychologically based, even if socially enculturated. We shall be describing how these triggers and resonances can be enhanced and sensitised.

### 3.3. *Confirmation of knowing*

The educational institutionalisation of knowledge as an object rather than as a state of awareness (as with Russell and some Eastern philosophies), supports inert rather than active knowledge through the development of forms of justification and confirmation of knowing.

Confirmation of *knowing-that* begins for young children in external authority, and during schooling one hopes that it turns into personal reasoning. Similarly, sources of confirmation for *knowing-how* also begin with external authority of the community, as practices are picked up through participation (peripheral or central), but one hopes that this begins to be augmented and displaced by personal judgement of appropriateness or fit. Sources of confirmation for *knowing-why* likewise begin with external authority of the stories that adults weave, but one hopes that these are displaced by a sense of personal and independent coherence.

Thus ideal education can be seen as a process of moving from external to internal authority, whereas in practice, with rapidly expanding student numbers, education is driven to training of explicitly stated behaviour while maintaining external authority. Mathematics has traditionally been seen as a refuge for the preservation of external authority, yet becoming a mathematician involves refining justifications to convince oneself, to convince a critical friend, and to convince a reasonable sceptic (Mason et al., 1982).

Confirmation of knowing-why can be placed within structure manifested in symbols, and structure manifested within images. Thus one of the differences proposed between Bohr's approach to physics and Heisenberg's was that Heisenberg was satisfied if the mathematics 'worked', because for him the mathematics 'was the explanation, the why'. Bohr wanted to be able to explain the physics in simple words to a general citizen (Frayn, 1998). Both sought confirmation of knowing-why, but for one that confirmation lay within symbols, and for the other, within images. This distinction is refracted into a difference between students of engineering, science, computer science, and economics who are forced to study mathematics because 'they will need it' and simply want it 'to work' (justification lies in getting the right answers), and those who study mathematics for itself (justification lies within structure expressed in symbols).

*Knowing* which is confirmed only in external sources is not only much less robust, but also less likely to be brought to mind than knowing which can be reconstructed from structural elements, since it is those structural elements which are most likely to be triggered by novel situations. By contrast, sources of confirmation of *knowing-to* act begin and stay with personal judgement of appositeness of what came to mind, in the light of the consequences of actions, augmented by external suggestions for modification, praise, or criticism, of those actions.

## 4. NOT KNOWING-TO ACT

It is impossible to be aware in the moment that you do not know-to act, but you can become aware afterwards. Teachers recognise when students 'don't know to act', because their own awareness extends beyond the awareness of their students, but they themselves often suffer not knowing to act while teaching.

Schoenfeld (1988) describes in detail how students who were known to know all requisite geometrical facts did not think to employ them to achieve a geometric construction. Hoyles (1998) describes a group of teachers who had resolved a construction problem involving using dynamic geometry software, a special case of which would prove that a certain configuration would be impossible. When asked to construct that configuration, they all started by trying to make a construction, and upon failing, only then realised that it was impossible. One way to account for the teachers not knowing to apply their previous result is that they had established a way of working using the software, and it was only during the use of the tool that they really encountered the conditions of the problem and recognised their recent experience. Alternatively, they were locked into succeeding with construction tasks, and were not expecting an impossible one.

In a similar vein but with regard to teaching rather than doing mathematics, Gates (1993) describes a novice teacher going into a classroom with a single object that he wanted each student to handle before proceeding with the lesson. Of course, it took a long time for the object to pass around the classroom, disrupting his lesson plan which did not take account of this. Asked about it later, he replied 'I just didn't think of it'. Spence (1996) reports in detail on two teachers and the sorts of actions they knew and did not know-to carry out in the midst of their teaching despite ample evidence of knowing-about those acts as possibilities before hand.

In his play about the visit of Heisenberg to Bohr in 1941, Frayn (1998) offers many interpretations of the visit, but ends with the possibility that Heisenberg failed to think of performing a particular mathematical calculation, and that if he had done, it would have resulted in him knowing-how to construct a hydrogen bomb. Not knowing-to calculate, in the moment, might have meant a rather different outcome of the second world war.

## 5. KNOWING-TO ACT IN THE MOMENT

The purpose of distinguishing *knowing-to* from other kinds of knowing is that it is precisely the absence of knowing-to which blocks students and teachers from responding creatively in the moment. Whereas de Jong and Ferguson-Hessler (*op cit.*) approach what we call knowing-to by analysing other people's knowledge-in-use, we want to get at the lived experience of how it is that one knows-to act in the moment, in order to develop ways to learn from experience, to work on sensitising oneself to know-to act with a broader range of responses in the future. We offer three approaches to what we mean by *knowing-to* act in the moment:

(i) an experiential approach, which in text emerges as a collection of descriptions of incidents drawn from teaching and doing mathematics;
(ii) a more theoretical yet also experience-based description of distinctions developed from the preceding sections but dwelling on knowing-to; and
(iii) a more theoretical approach intended to illuminate what we mean by *knowing-to* within the learning and doing of mathematics.

### 5.1. *Experiential approach*

Since it is inappropriate here to try to generate awareness of not-knowing in you the reader, we offer instead some descriptions of incidents. We hope that you will recognise the essence of each in your own experience.

Student errors:

Students well versed in solving equations, and well aware that when both sides of an inequality are multiplied by the same negative quantity, the inequality sign reverses, nevertheless when faced with the inequality $\frac{1}{x-1} > 1$, multiplied both sides by $x - 1$ without considering the implications of that factor being negative (Barrington, private communication).

One way to account for students not knowing in the moment to check the sign of their multiplying factor is that they require full attention to the calculation in hand, so there is insufficient free attention to attend to the result of any metonymic trigger or metaphoric resonance between multiplying and inequalities. Another explanation is that working with equalities tends to stress comparing things by division, and students are much less aware that taking a difference is also a good strategy for comparing algebraic quantities, as it avoids the error described here.

JM:

During a workshop one participant asked if they had reached a correct conclusion, and I answered immediately with a 'yes'. In later reflection on the session, participants were critical of my response, suggesting that a provocative 'why do you want re-assurance?' could have been more effective. Although I frequently engage in such meta-cognitive responses, in this instance it never occurred to me. I did not notice an opportunity. I did not know-to make a metacognitive shift, despite having written about using such responses.

There are a number of extenuating factors which might account for not-knowing, but putting these forward would detract from the core of the experience, which is realising later that he had been tunnel-visioned in the moment.

MS:

As a child I was passionate about insects. I did graduate work immersed in studying bugs and as a science undergraduate I passed courses in geology. As a teacher of all sciences years 6 through 8, my teaching about insects 'flew', whereas my teaching of geology just 'deposited' facts. I pre-planned lessons in each topic, but whereas for geology we stuck to my plan, with the insects things changed frequently, even in mid lesson. We laughed a lot, there was excitement and energy as we studied insects in the class, and by the end of each year, most students were confident in handling them. The students seemed to share my appreciation. By contrast, although I know facts about rocks, they don't speak to me. So in the classroom, although there were plenty of samples and time was spent handling, exploring and classifying, I presented them to the students and tested them on their knowledge. At the end of the topic I was relieved that we could move on to something else.

One way to account for this difference is that with the insects she was genuinely interested herself. It tripped off her tongue to say 'I don't know …how can we find out?' and to explore with the students, whereas with the minerals, she was hesitant to admit to not knowing. She would go and find out as an obligation to the students, but she lacked the confidence to explore publicly with students. She was not confident that she knew to act in useful ways spontaneously.

### 5.2. *Beyond knowing-about*

Experiences such as those reported above, led us to link our distinctions diagrammatically with those of Ryle:

*knowing that* something (factual) is true;
*knowing how* to do something, how to carry out some procedure;
*knowing why* in the sense of having some stories I tell myself to account for something;
*knowing to* act in the moment.

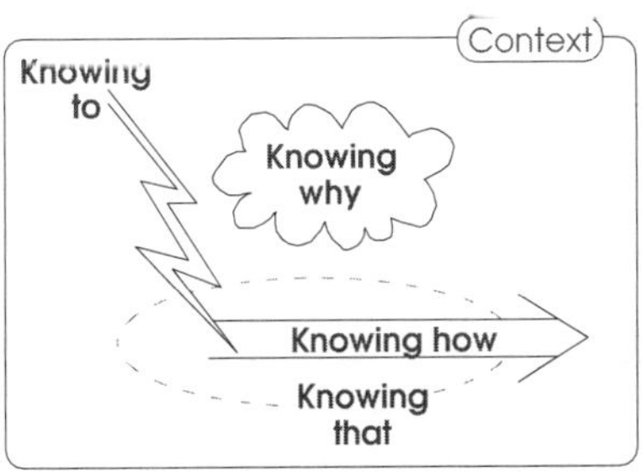


*Figure 1.*

We see knowing-to as instantaneous, a gestalt. It is either seamlessly present and so un-noticed, or else it arrives suddenly like a bolt of lightning, a moment of hazard that briefly illuminates, like the raven-spirit in Haida creation stories (Reid, 1980). Another way to describe it is in terms of the common experience of looking in a crowd for someone you know well: you don't actually scan for their features, rather you scan generally and let their image 'find you'; suddenly they stand out from the rest of the crowd.

Knowing-to is often marked by a sudden shift of the focus or locus of attention. To verify this, you have to try to notice a moment when an idea comes to you. This is difficult because your attention is likely to be fully caught up by the arriving idea! The fact that ideas come suddenly (even if there is a long period of preparation) follows from the observation that there is a sudden shift in the structure of attention, and such shifts are instantaneous.

For example, while pondering (gazing at) a geometrical figure, the impulse arises to draw in a line-segment or other element, and suddenly there is more to work with overtly. In what sense does one 'know-to' add that element? Sometimes you can trace back to a desire to compare some angles or segments that were not physically present, so the added element is merely the manifestation of an imagined one. But often the thought arises suddenly, like a bolt from the blue. Similarly, faced with a mess of algebraic symbols, suddenly a partial pattern catches the eye; not the sort that will enable a symbol processor to factor the whole expression, but still a recurring pattern of symbols. Knowing-to collapse those to a single symbol may enable other patterns to emerge. The pattern and the substitution are examples of knowing-to, the one arising unexpectedly, the other, as a result of experience.

More prosaically it is one thing to know-how to add or multiply numbers, but quite another to know which operation to use in a given context; one thing to know-how to calculate a derivative or a greatest common divisor, but quite another to think of doing it without direct cues. Brown (1981) captured this beautifully in her title 'Is It an Add or a Multiply, Miss?' which is played out in primary, secondary and tertiary classrooms all over the world.

Once the moment of knowing-to takes place, knowing-how takes over to exploit the fresh idea; knowing-that forms the ground, the base energy upon which all else depends and on which actions depend; knowing-why provides an overview and sense of direction that supports connection and link making and assists reconstruction and modification if difficulties arise en route. Knowing-how provides action, things to do, changing the situation and transforming it, and providing the various knowings with fresh situations upon which to operate. All this takes place within a context-environment. The state of sensitivity-awareness of the individual, combined with elements of the situation which metonymically trigger or metaphorically resonate with experience, are what produce the sudden knowing-to act in the moment.

Awareness of knowing and of not knowing is crucial to successful mathematical thinking. Most people have been stuck on a problem, and vaguely aware of being stuck, but as long as their awareness of their state remains below the surface of action they are unable to act upon that awareness by intentionally invoking some strategy. They continue to be stuck, staring at a blank page or repeating the same calculations over and over hoping for a break-through. Bauersfeld (1993) refers to the experience of repeating what you *can* do in the absence of being able to do anything else, and story-tellers capture it with the paradox of a person at night searching for their door key under the street light and not where they dropped it in front of their door.

Mary Boole (Tahta, 1972) made use of awareness of not-knowing to suggest that the origins of algebra lie in *acknowledging ignorance* of the answer to a problem. Knowing you do not know enables you to denote what you do not know by some symbol, and then to treat it as if it were known in order to write down expressions or properties, eventually arriving at knowledge of what previously you were ignorant. The key importance of such awareness is reflected in a Sarmoun recital (Shah, 1968):

> ... He who does not wish to know, and yet says that he needs to know: let him be guided to safety and to light.
> He who does not know, and knows that he does not know: let him, through this knowledge, know.
> He who does not know, but thinks that he knows: set him free from confusion and ignorance. ... (p. 253)

A possible lesson from this ancient wisdom is that not-knowing is a valuable and even crucial state, because from it, knowing can follow. But it may require acknowledging the fact of not knowing in order for other prepared strategies to then come to mind. Much good advice offered to students goes unheeded through the student not recognising that they do not know.

In order to know-to draw student attention to meta aspects such as how they are working and how they might work when studying, an extra degree of awareness is required on the part of the teacher: knowing-to as well as knowing-how to create suitable conditions and then to direct student attention effectively. Strategies come most readily to mind if they are richly linked with past experiences. A tutor can assist this by providing conditions in which students get stuck, and then come to recognise their use of powers to get themselves unstuck, and come to recognise the scaffolding–

fading employed by the tutor. If problem-solving becomes an object of instruction, the inevitable forces of *transposition didactique* and *contrat didactique* are likely to subvert the tutor's intentions.

## 5.3. *Theoretical approach*

Another way to approach the notion of knowing-to is through distinctions arising from the triple of terms *noticing–marking–recording* (Mason, 1996). If you notice and mark something, then you are in a position to initiate a re-mark about it to someone else. By contrast, you may recognise what someone else is talking about, but not have been able to initiate the remark yourself: this is noticing without marking. Finally, you may not only mark and re-mark, but actually choose to make a record in some re-accessible form.

These distinctions provide a way of thinking about knowing-in-the-moment:

*responsive–knowing* (noticed but not marked) which produces background experience through which to respond in future;
*active–knowing*, in which you can actively initiate use of knowing (noticed and marked);
*reflective* or *meta-knowing*, in which you are aware of knowing and can describe and reflect upon that knowing (noticed, marked, and recorded).

For example, when a teacher finds that they have to remind students of some facts or techniques or topics in order that they can do a particular problem, the students exhibit *responsive-knowing* which is not sufficiently active to come to mind spontaneously. When a student knows to differentiate in order to locate an extremum (and knows to check that the function is differentiable etc. and to examine values at end-points), then the knowing is active; they *know-to* act without having extra cues. When a student can describe in general terms 'how to do such a question' and can make up easy, hard and general questions of that type, they have reflective-knowing and are perhaps ready to consider how to psychologise the subject matter (Dewey, 1902) for others and hence begin to teach others. The extent to which they need cues in the task statement or explicit scaffolding from an relative expert, provides evidence of the intensity and complexity of their knowing-to.

The common refrains referred to in the introduction are sometimes accounted for by using the notions of instrumental and relational understanding (Skemp, 1979).[1] Relational understanding comprising knowing-that, -how, and -why implies and imputes that the student is able to generate actions from understanding (their knowledge is 'ert'), whereas when understanding is instrumental, students are at the mercy of what comes to mind in the moment (knowledge is inert and triggered haphazardly). But even when people are thought to understand thoroughly, possibilities to act may not necessarily come to mind at the crucial moment (*viz.* Heisenberg). Evidence for this is provided by the all too frequent retrospective wish 'If only I had thought to (known to) . . . '.

In other words, where awareness is assumed to have been educated because behaviour appears to have been trained, students may in fact be much less flexible than expected, suggesting that student concentration has actually been on behaviour rather than understanding. For example, generations of students have been taught fractions 'from the beginning', year after year, because they appear not to recall anything from the previous year. This is not at all surprising, since when students experience difficulty, they tend to seek assistance in the practical and immediate problem of training their behaviour (how do you do it?) rather than in educating their awareness (what is it about?). Some students are content to get answers to immediate tasks and move on, while others want to feel confident that they could get the answer to another similar problem in the future. These latter students are often dismissed or short-changed in teachers' desire to be pragmatic and 'get students through the test'. Even when students ask 'Why are we doing this?' or 'What is it all really about' they are frequently signalling not a desire for relevance so much as loss of confidence in performance.

From an educator's point of view, the trouble with distinctions between kinds of knowing, such as Ryle's, is that although they are useful for highlighting one's own experience, they do not provide categories for making observations of others. To see that *knowing that, how*, and *why* are not readily observable, we note that:

*Knowing that* can be achieved instrumentally
by memorising mnemonics such as SOHCAHTOA for the trigonometric ratios or CAST for the signs of the trig ratios in different quadrants;

by habituation through sufficient repetition and use, such as learning tables, or 'minus times minus is plus', or where the minimum occurs for a quadratic.

*Knowing that* can be achieved relationally
by connection and image: the sine of an angle is what it is (i.e. not known through a formula or mnemonic but perceived directly);

*Knowing how* can be achieved instrumentally
by memorising a procedure or formula, complete with inner 'incantations' that support the procedure, like the quadratic formula, or the triangle mnemonic for velocity $\frac{d}{v|t}$;

*Knowing how* can be achieved relationally
through reconstructing: distance = speed $*$ time from the units in which each is measured; trigonometric multiple angle formulae from an awareness of how they are generated.

*Knowing why* can appear to be achieved instrumentally
through memorising: definitions of limit, continuity, area, . . .

*Knowing why* can be achieved relationally
through constructing your own justifications and stories for what one is doing and why, such as what a limit is and how you justify one, what area is and how you find it, etc.. It helps to know that other people also have such stories, and to hear those stories, but each persons stories are their own.

Thus, the fact of observable behaviour from students is insufficient to gauge the kind of knowing, for it is only when put in a situation that can be resolved by a particular strategy that you find out whether that strategy comes to mind, and sometimes an alternative strategy works just as well.

## 6. Situatedness of knowing-to

Education has traditionally isolated important skills (counting, arithmetic, algebra, study skills, etc.), abstracted them or turned them into procedures, and then taught these to young people. The intention is that students should then apply their skills in a variety of contexts. The trouble is that the traditional sequence of example, theory, exercise, application tends to leave the knowledge inert.

Situated cognition arose as a construct (Lave, 1988; Brown et al., 1989) when people questioned whether context-independent skills are possible, arguing that all skills are learned in context and that the situation is integral to changes in cognition that take place. Noss and Hoyles (1996, p. 105) highlighted situated abstraction as well, arguing that even when someone abstracts, they do so in a context of experienced variation. Marton and Fazey (preprint) go so far as to suggest that learning consists of extending the range of variability within which some aspect remains invariant, a decidedly mathematical formulation. Knowing about situatedness, and having a label for it provides helpful support for acting upon that knowledge, but they are insufficient in themselves to enable a teacher to act in the moment.

Independently of perspective and philosophy adopted about the central aims of education, there are fundamental problems associated with the notion of using or applying knowledge:

When education is dominated by abstraction and generalisation, there is an issue of application: how do students know to apply that knowledge in a given situation or context?

When education is dominated by training in behaviour, there is an issue of transfer: what enables a student who has mastered a skill to know to apply that skill in a novel context, and to generalise it? (Detterman and Sternberg, 1993).

When education is dominated by specific practical situations and contexts, there is an issue of generalisation: how do students come to stress and ignore appropriately, since they may stress features that are irrelevant and ignore features that are relevant?

When education is dominated by apprenticeship-like participation in a community of practice, so that training and education are context-dependently situated, there is an issue about extending contexts: how do experts and novices come to recognise similarity between contexts which enables them to transcend the situation in which they encountered an idea or topic, and employ it in a new context?

No matter how it is phrased, the issue reduces to how it is that in some situation or context, we *know-to* act in some way.

## 7. Knowing and Attention

We have already suggested that knowing-to requires more than trained behaviour. It requires some degree of sensitivity to features of a situation, some degree of awareness in the moment. In this section we develop details of that connection.

Knowing in the moment is a state of readiness as a result of what is being attended to. It consists of what is primed and ready to come to attention, and it excludes what is blocked or otherwise unavailable. What is the student attending to at any given moment? What is the structure of their attention? What is blocked from their view? Is their attention tightly focused on some aspect or detail? Is it multiple rather than single? Is it diffuse or clear? Is it flexible or rigid? Are they caught by detail or is part of their attention available to observe themselves? Can students be helped to be aware of their attention, of being stuck or of being caught by habits?

In Mason and Davis (1988) and Mason (1989, 1998) it is suggested that coming to know is essentially a matter of shifts in the structure of attention, in what is attended to, in what is stressed and what consequently ignored with what connections. In other words, it has to do with whether attention is uni-focal or multi-focal, and with the degree of breadth and flexibility of that attention. Knowing is not a simple matter of accumulation. It is rather a state of awareness, of preparedness to see in the moment. That is why it is so vital for students to have the opportunity to be in the presence of someone who is aware of the awarenesses that constitute their mathematical 'seeing'.

### 7.1. *Shifts of attention*

How shifts occur in what someone attends to as a result of a particular stimulus in a particular context, is of fundamental importance to anyone interested in epistemology, for it is only when the very structure of attention changes that learning could be said to have taken place. Three mechanisms are frequently discussed and are summarised here.

*Habituation and enculturation*

Habituation (Mason and Davis, 1989) is a fundamental process through which people come to know. Sometimes it is an enculturation through participation in a cultural practice (you learn to prove things mathematically by seeing many proofs go past you); sometimes it is a process in which you suddenly find that you have changed your perception ($.9999\ldots = 1$ in the standard reals is often experienced this way), especially when you are called upon to teach it. Habituation is highly problematic: it seems to require some sort focused attention; it can arise through mere immersion in a social practice, but it seems to require some emotional-affective component.

*Generality and particularity*

Coming to see particulars as particular examples *of* a generality, and to see particular instances *in* or *of* the general involves a shift of attention, from the item itself to the item as representative. Mathematical structure is about being aware of relationships, and these are generalised and abstracted from particulars, yet still tied to those particulars through re-constructability. The spiral *manipulating–getting-a-sense-of–articulating* (Mason, 1993) based on Bruner's triad of *enactive–iconic–symbolic* (Bruner, 1966) is but one way to express this. Michener (1978), Polya (1962), Krutetskii (1976), Davidov (1990) and many other authors have underlined the centrality of generalisation in mathematics, yet generality nevertheless seems not to be fully appreciated by many teachers and educators.

*Metonymic triggers and metaphoric structures*

Playful connectedness experienced through subconscious links triggered by something like a Proustian 'clink' provides a rich but largely idiosyncratic web of meaning for each individual. Analogies and images (equations as balances, numbers as bundles of tens and hundreds, integrals as continuous summations, multiplication by –1 as a rotation through a half turn) form ways of perceiving which can be deeply embedded as frozen metaphors (for example, 'getting through the material for this lecture') or consciously invoked ('if I think of a square number as the area of a square then . . . ' ). Metonyms, metaphors, and images cannot be implanted intentionally with uniform success: they seem to have to be taken up or activated by the individual.

## 8. How knowing-to can prepared for

*Knowing-to* is developed through connections being established between past, present, and future, so that in the future, past experience in-forms (literally) practice in the moment. In order to inform practice, it is essential that something brings to mind a possible action in the moment, just in time when it is needed, and not in retrospect, as captured by the French idiom *l'esprit d'escalier*: the good idea about what one could have done, that comes too late, after the event. It is a commonplace that we learn from experience, but in fact one thing we seem not to learn from experience is that we rarely learn from experience alone. Some action, some preparation for the future is required. The tricky bit is moving the moment of recognition of a possibility from the retrospective to the present.

### 8.1. *Reflection*

To bring a possibility to mind in the moment requires either luck (as in the Raven metaphor mentioned earlier) or intentional preparation. Reflection is usually put forward as the sort of preparation required, but the term *reflection* has become too broad and diffuse in meaning to carry significance in itself. What is required is active re-vivifying of recent incidents or moments, coupled with actively imagining oneself in a similar situation in the future in which you act as you would like. Thus you need to collect alternative actions which you want to try, and you need to develop sensitivity to noticing opportunities in which to try them.

*Reflection-on-action* and *reflection-in-action* (Schön, 1983, 1987), *looking back* (Polya, 1945), *reflective abstraction* (Piaget, 1977; Dubinsky and Levin, 1986) are all descriptions of aspects of a process of forming repeatable connections to inform future practice by altering the form and structure of attention, in order to develop *knowing-to*. Collins et al. (1989) gave three case studies of cognitive apprenticeship, one of which is based on Alan Schoenfeld's sessions with students, in which he employs a variety of reflective practices.

Turning these sentiments-about into practical action is the underlying approach taken in Mason et al. (1982), and in materials produced in the Centre for Mathematics Education at the Open University over 15 years. *The discipline of noticing* (Mason, 1996) elaborates and justifies the approach epistemologically and methodologically, promoting reflection-in-action through various forms of reflection-on-action.

One issue for people employing such practices repeatedly is that frequent repetition of the same prompts for reflection are likely to turn interest into mechanicality (Northfield and Baird, 1992). It is vital to keep the reflective process fresh in order that its effects be fresh for students.

### 8.2. *Mechanisms for supporting knowing-to*

Experience with Open University students, who study at a distance from texts, has shown that students can be helped by provoking them to use fundamental powers such as specialising and generalising, and imagining and expressing, and then bringing to their attention the fact that these powers are available to them to use when they become aware of being stuck. We have also found it useful to suggest questions for students to ask themselves:

*What do I know?* (which takes very little time to write down or mentally review when working on a specific problem)
*What do I want?* (which focuses attention on the current problem, which may be a subset of the original) (Mason et al., 1982)

If these also fail to suggest something, then students can work through the technical terms and re-write them in ordinary language, then look for theorems that connect them in some way.

The crucial aspect of supporting others in fostering and sustaining their mathematical thinking lies in helping them recognise and accept being stuck in the first place, without criticism, and then to recognise having an idea come to mind when it does come (knowing-to act).

Knowing in terms of *having connections* is richest when there are multiple links that may be activated intentionally or subconsciously, or as Hiebert and Carpenter (1992), Noss and Hoyles (1996), Burton (in press) and others before and since have put it, when a rich *web of meaning* has been constructed. The more senses engaged, the more dimensions of the psyche positively awake, the richer are the possibilities. In the moment it does not matter what connections someone has been exposed to or has recognised in the past. What matters is what they are aware of in that moment, what can occupy their attention and how that attention is structured.

There are two linguistic mechanisms which bring things to mind, and which have been mentioned in passing already: metonymic triggers and metaphoric resonances. Metonymic triggers are associations and affective connections that take place below the surface of consciousness, tripping playfully along a chain of signifiers (Lacan, 1985). Metaphoric resonances include insights based upon analogies and similes, and are based on stressing some aspects (structure) while ignoring others (detail). Thus terms such as *differential equation*, and *linearity*, are metonymic in that they describe an object by one or two of its aspects, yet sometimes their use is to invoke a corresponding metaphor (Pimm, 1987). Positive metonymic links (as distinct from negative emotions) cannot be deliberately established; rather, they arise from a playful attitude and they depend on a degree of personal confidence. Metaphoric resonances can be supported (but not guaranteed) by attending to the semantics, by becoming aware of structure. Periods of 'looking back', of reflecting upon what has been noticed certainly contribute to the awareness needed to resonate in the future. Both types of action can be developed through the use of labels.

## 8.3. *Labels*

The most powerful mechanism available to teacher and student alike is labelling. A label, or framework, is a succinct collection of words, most powerfully a triple, which act as an axis around which experiences can collect. If the label words are likely to arise in the course of an unfolding situation (for example, *general, particular, special*), then there can be a direct metonymic trigger into a slogan such as *seeing the general through the particular, and the particular in the general*. If such a slogan has been used to summarise experiences the student has had, and if it has been linked with specific actions which might be taken when doing mathematics, or when working with students on mathematics, then the natural sense-making powers of the individual have the data from which to generate knowing-to in the moment.

The important factors are to provoke students into actions, to draw those actions to their attention, to use or negotiate some label for those actions, and then to use that label, but less and less directly, to remind (literally) students of possible actions. As the prompts become less and less direct, the students take on the initiative, so that they know-to employ those strategies and actions themselves. A useful meta-framework for remembering this as a teacher is *directed–prompted–spontaneous* or *scaffolding–and–fading*.

The techniques incorporated by the *discipline of noticing* (*op cit.*), drawn from many sources, provide methods for teachers to work on knowing-to act freshly in the midst of their teaching, and thereby to be able to support students in knowing-to respond creatively and thoughtfully to novel tasks, based on educating awareness and training behaviour through harnessing of emotions. Where only one dimension is activated, where past and present experience are juxtaposed without intention, learning is attenuated, and knowing depends on the strength of random connections. Where the individual's intention is the driving force to link past and present experience, rich networks of associations provide the basis for future resonance.

Anything with positive potential also has negative potential, and labels and slogans are no exception. Slogans can become superficial jargon at least as easily as they can become nodes for intricate webs of personal connections. Mnemonics intended to serve students as labels often block effective subordination as well as triggering recall. For example, an established mnemonic like the CAST rule for signs of trigonometric functions can force you to work through the signifying mnemonic each time rather than moving directly to the signified, even when frequent use is made of the rule.

## 9. Practice and Attention

One of the common educational confusions is to assume that the best, perhaps only way to know-to do something is to gain facility through 'practice until perfect'. Certainly gaining facility reduces the amount and simplifies the structure of attention required, and certainly if there is a choice of tools, people always choose the tool with which they are most familiar. Yet we all have experience that shows that practice to mastery is insufficient in itself. Many techniques have been rehearsed until perfect, only to fail to be employed when appropriate or needed, while some techniques are learned without ever actually practicing at all.

Hewitt (1994) has explored in detail Gattegno's notion (Gattegno, 1987) that integration through subordination is how human beings develop skills and associated knowing-to, so that tasks which provoke rehearsal while diverting as much attention as possible *away* from that rehearsal are much more likely to be effective than practice that reinforces the employment of full attention on the skill being practiced. For example, in order to gain facility at multiplying matrices or adding fractions or being able to say that 3 is four less than 7, it is vital to have surface attention attracted towards some goal such as a game or an exploration, while many calculations of the desired type are required for the 'larger' purpose.

## Conclusion

It is our contention that once you become aware that knowing-to act is not the same as possessing knowledge, you can direct your attention to the psychological and sociological factors that are at play. You can work at your own knowing-to act by intentionally making fresh choices in teaching, in order to sensitise yourself to techniques that you have for preparing yourself to know-to act in the future. Fundamental among these techniques is the use of mental imagery for pre-paring, that is for mentally imagining a future situation as vividly as possible, and imagining yourself doing whatever it is you want to think to do. Students who have successfully employed some process such as specialising or generalising or working backwards etc., can gain most from their success by mentally re-entering the situation as best they recall it, and re-experiencing something of what occurred to them, then mentally imagining themselves in a similar situation (and this is where what is worth stressing and what worth ignoring can be negotiated between several students and/or the teacher) and carrying out a similar act.

Simple as it sounds, it takes considerable effort. It requires work to move the moment of noticing an opportunity to act from the retrospective into the moment. The energy required for that effort is available from reflection immediately after some period that includes a successful incident, but since it dissipates rapidly, reflection is much less effective if delayed for hours or days.

## Notes

[1] Mercedes McGowen (1998) points out that although the distinction between relational and instrumental understanding is usually attributed to Skemp, he acknowledged that it was a remark of Mellin-Olsen which set him off along that track.

## Acknowledgements

We are most grateful to the editor, Dina Tirosh for her faith in the substance of the paper and her continued encouragement, and to the referees for repeatedly making positive and useful comments which have improved the presentation significantly.

## References

Bauersfeld, H.: 1993, 'Theoretical perspectives on interaction in the mathematics classroom', in R. Biehler, R. Scholz, R. Sträßer and B. Winkelmann (eds.), *The Didactics of Mathematics as a Scientific Discipline*, Kluwer, Dordrecht, The Netherlands.

Bereiter, C. and Scardammalia, M.: 1989, *Knowing Learning and Instruction: Essays in Honor of Robert Glaser*, Lawrence Erlbaum, Hillsdale.

Biggs, J.: 1994, 'Modes of learning, forms of knowing, and ways of schooling', in A. Demetriou, M. Shayer and A. Efklides (eds.), *Neo-Piagetian Theories of Cognitive Development*, Routledge, London, pp. 31–51.

Brousseau, G.: 1984, 'The crucial role of the didactical contract in the analysis and construction of situations in teaching and learning mathematics', in H. Steiner (ed.), *Theory of Mathematics Education*, Paper 54, Institut fur Didaktik der Mathematik der Universität Bielefeld, pp. 110–119.

Brousseau, G.: 1997, *Theory of Didactical Situations in Mathematics: Didactiques des Mathématiques, 1970–1990,* (translated by N. Balacheff, M. Cooper, R. Sutherland, V. Warfield), Kluwer, Dordrecht, The Netherlands.

Brown, J., Collins, A. and Duguid, P.: 1989, 'Situated cognition and the culture of learning', *Educational Researcher* 18(1), 32–41.

Brown, M.: 1981, 'Is it an add, or a multiply, Miss?' (part 3), *Mathematics in Schools* 10(1), 26–28.

Bruner, J.: 1966, *Towards a Theory of Instruction*, Harvard University Press, Cambridge.

Burton, L.: 1995, 'Moving towards a feminist epistemology of mathematics', *Educational Studies in Mathematics* 28(3), 275–291.

Burton, L. (ed.): (in press), *Learning Mathematics: From Hierarchies to Networks*, Falmer Press, London.

Chevellard, Y.: 1985, *La Transposition Didactique,* La Pensée Sauvage, Grenoble.

Clancey, W.: 1993, 'Situated Action: a neuropsychological interpretation response to Vera & Simon', *Cognitive Science* 17, 87–116.

Collins, A., Brown, J. and Newman, S.: 1989, 'Cognitive apprenticeship: teaching the crafts of reading, writing, and mathematics', in L. Resnick (ed.), *Knowing, Learning, and Instruction: essays in honor of Robert Glaser,* Lawrence Erlbaum, Hillsdale, pp. 453–494.

Davidov, D.: 1990, 'Types of Generalisation in Instruction', *Soviet Studies in Mathematics Education* Vol. 2, NCTM, Reston.

Detterman D. and Sternberg, R. (eds.): 1993, *Transfer on Trial: Intelligence, Cognition, and Instruction*, Ablex, Norwood N.J.

De Jong, T. and Ferguson-Hessler, M.: 1996, 'Types and qualities of knowledge', *Educational Psychologist* 31(2), 105–113.

Dewey, J.: 1902, *The Child and the Curriculum*, U of Chicago Press, Chicago.

Dubinsky E. and Levin P.: 1986, 'Reflective abstraction and mathematics education: the genetic decomposition of induction and compactness', *Journal of Mathematical Behaviour* 5, 55–92.

Fennema, E. and Franke, M.: 1992, 'Teachers' knowledge and its impact', in D. Grouws (ed.), *Handbook of Research on Mathematics Teaching,* MacMillan, New York, pp. 147–164.

Frayn, M.: 1998, *Copenhagen*, Methuen Drama, Methuen, London.

Gates, P.: 1993, ' "I just didn't think of it": learning to teach mathematics', *Proceedings of the British Society for Research in Learning Mathematics Meeting May 22*, pp. 29–34.

Gattegno, C.: 1987, *The Science of Education Part I: Theoretical Considerations*, Educational Solutions, New York.

Greeno, J. Smith, D. and Moore, J.: 1993, 'Transfer of situated learning', in D. Detterman and R. Sternberg (eds.), *Transfer on Trial: Intelligence, Cognition, and Instruction*, Abbex, Norwood, NJ, pp. 99–167.

Harrison, G. (ed.): 1937, *A Book of English Poetry*, Penguin, Harmondsworth.

Hewitt, D.: 1994, *The Principle of Economy in the Learning and Teaching of Mathematics*, unpublished PhD dissertation, Open University, Milton Keynes.

Hiebert, J. and Carpenter, T.: 1992, 'Learning and teaching with understanding', in D. Grouws (ed.), *Handbook of Research on Mathematics Teaching and Learning*, MacMillan, New York, pp. 65–93.

Hofer, B. and Pintrich, P.: 1997, 'The development of epistemological theories: beliefs about knowledge and knowing and their relation to learning', *Review Of Educational research* 67(1), 88–140.

Honderich, T.: 1995, *The Oxford Companion to Philosophy*, Oxford University Press, Oxford.

Hoyles, C.: 1998, 'A culture of proving in school maths?', in D. Tinsley and D. Johnson (eds.), *Information and Communications Technologies in School Mathematics*, Chapman & Hall, London, pp. 169–182.

Kang, W. and Kilpatrick, J.: 1992, 'Didactic transposition in mathematics textbooks', *For The Learning of Mathematics* 12(1), 2–7.

Kieren, T.: 1994, 'Bonuses of understanding mathematical understanding', in D. Robitaille, D. Wheeler and C. Kieran (eds.), *Selected Lectures from the 7th International Congress on Mathematical Education*, Les Presses de l'Université Laval, Quebec, pp. 211–228.

Krutetskii, V.: 1976, *The Psychology of Mathematical Abilities in Schoolchildren*, University of Chicago Press, Chicago.

Lacan, J.: 1985, 'Sign, symbol and imagery', in Blonsky (ed.), *On Signs*, Blackwell, Oxford.

Lave, J.: 1988, *Cognition in Practice: Mind, Mathematics and Culture in Everyday Life*, Cambridge University Press, Cambridge.

Locke, J.: 1894, (Dover reprint 1959): *An Essay Concerning Human Understanding*, 2 Vols, Dover, London.

Marton, F. and Fazey, J.: (preprint), *Understanding as Space of Experienced Variation.*

Mason, J., Burton, L. and Stacey, K.: 1982, *Thinking Mathematically*, Addison Wesley, London.

Mason, J. and Davis, J.: 1988, 'Cognitive and metacognitive shifts', *PME XII*, (ed. Borbás) Vol. 2, pp. 487–494.

Mason, J. and Davis, J.: 1989, 'The inner teacher, the didactic tension, and shifts of attention', in G. Vergnaud, M. Rogalski and M. Artigue (eds.), *Proceedings of PME XIII*, Paris, Vol. 2, pp. 274–281.

Mason, J. and Spence, M.: 1998, 'Towards a psychology of knowing-to', in C. Kane, M. Goos and E. Warren (eds.), *Proceedings of MERGA* 21, Coolangatta, New South Wales, Australia.

Mason, J.: 1989, 'Mathematical abstraction seen as a delicate shift of attention', *For the Learning of Mathematics* 9(2). 2–8.

Mason, J.: 1993, 'Working on awareness', in J. Searl (ed.), *Proceedings of the Edinburgh Mathematics Teaching Conference*, University of Edinburgh, Edinburgh.

Mason, J.: 1996, *Personal Enquiry: Moving from Concern towards Research*, Open University, Milton Keynes.

Mason, J.: 1998, 'Structure of attention in teaching mathematics, Plenary address to Canadian Mathematics Education Study Group 15th meeting, May', in Y. Pothier (ed.), *Proceedings of cmESG 15*, in preparation.

Maturana, H. and Varela, F.: 1972, 'Autopoesis and cognition: the realization of the living', D. Reidel, Dordrecht, The Netherlands.

Maturana, H. and Varela, F.: 1988, *The Tree of Knowledge: the Biological Roots of Human Understanding*, Shambala, Boston.

Maturana, H.: 1988, 'Reality: the search for objectivity or the quest for a compelling argument', *Irish Journal of Psychology* 9(1), 25–82.

McGowen, M.: 1998, 'Cognitive units, concept images, and cognitive collages', Unpublished PhD thesis, Warwick University, Coventry.

Michener, E.: 1978, 'Understanding understanding mathematics', *Cognitive Science* 2, 361–383.

Miller, L., Malone, J. and Kandl, T.: 1992, 'A study of secondary teachers' perceptions of the meaning of understanding', *AERA paper*, San Francisco.

Northfield, J. and Baird, J.: 1992, *Learning from the PEEL Experience*, Monash University Printing Service, Melbourne.

Noss, R. and Hoyles, C.: 1996, *Windows on Mathematical Meanings,* Mathematics Education Library, Kluwer, Dordrecht, The Netherlands.

Perry, W.: 1968, *Forms of Intellectual and Ethical Development in the College Years: a scheme*, Holt, Rhinehart & Winston, New York.

Piaget, J.: 1950, 'Introduction á l'epistemologie genetique', presses Univ. de France, Paris.

Piaget, J.: 1977, *The Development of Thought: Equilibration of Cognitive Structures*, (translated by A. Rosin), Harvard University Press, Cambridge.

Pimm, D.: 1987, *Speaking Mathematically*, Hodder & Stoughton, London.

Pirie, S. and Kieren, T.: 1994, 1Growth in mathematical understanding: how can we characterise it and how can we represent it?', *Educational Studies in Mathematics* 26(2–3), 165–190.

Polya, G.: 1945, *How To Solve It*, Princeton University Press, Cambridge.

Polya, G.: 1962, *Mathematical Discovery: On Understanding, Learning, and Teaching Problem Solving*, Combined edition, Wiley, New York.

Reid, B.: 1980, *The Haida Legend of The Raven and The First Humans*, Museum Note No. 8, U. B. C. Museum of Anthropology, Vancouver.

Renkl, A., Mandl, H. and Gruber, H.: 1996, 'Inert knowledge: analyses and remedies', *Educational Psychologist* 31(2), 115–121.

Russell, B.: 1914, *Our Knowledge of the External World as a Field for Scientific Mathod in Philosophy*, George Allen & Unwin, London.

Ryle, G.: 1949, *The Concept of Mind*, Hutchinson, London.

Schoenfeld, A.: 1988, 'When good teaching leads to bad results: The disasters of 'well taught' mathematics classes', *Educational Psychologist* 23, 145–166.

Schön, D.: 1983, *The Reflective Practitioner: How Professionals Think in Action*, Temple Smith, London.

Schön, D.: 1987, *Educating the Reflective Practitioner*, Jossey-Bass, London.

Shah, I.: 1968, *The Way of the Sufi*, Jonathon Cape, London.

Shulman, L.: 1987, 'Knowledge and teaching: Foundations of the new reform', *Harvard Educational Review* 57(1), 1–22.

Sierpinska, A.: 1994, *Understanding in Mathematics*, Falmer Press, London.

Skemp, R.: 1979, *Intelligence, Learning and Action*, Wiley, Chichester.

Spence, M.: 1996, 'Psychologising algebra: case studies of knowing in the moment', unpublished PhD thesis, Madison Wisconsin.

Tahta, D.: 1972, *A Boolean Anthology: Selected Writings of Mary Boole on Mathematics Education*, Association of Teachers of Mathematics, Derby.

van Hiele, P.: 1986, *Structure and Insight: A Theory of Mathematics Education*, Academic Press, Orlando.

Vergnaud, G.: 1981, 'Quelques orientations théoriques et méthodologiques des recherches Françaises en didactique des mathématiques', *Actes duVième Colloque de PME,* Vol. 2, pp. 7–17, Edition IMAG, Grenoble.
Whitehead, A.: 1932, *The Aims of Education and Other Essays*, Williams & Norgate, London.

JOHN MASON and MARY SPENCE
*Centre for Mathematics Education,*
*Open University, Walton Hall,*
*Milton Keynes MK7 6AA,*
*U.K.*

THOMAS J. COONEY

# CONCEPTUALIZING TEACHERS' WAYS OF KNOWING

ABSTRACT. This article addresses issues related to the ways teachers learn mathematics and the teaching of mathematics and the relevance of those ways to their professional development. Preservice teachers' understanding of school mathematics lacks sophistication, a situation that needs to be addressed in mathematics teacher education programs. What is critical is the means by which they encounter and explore the mathematics they will be teaching. Fundamentally, their mathematical experiences need to be congruous with the kind of teaching we would expect of a reflective, adaptive teacher. The article contains both practical and theoretical considerations of how these experiences might be structured. Theoretical orientations for conceptualizing teachers' belief structures are offered as a foundation for conceptualizing teachers' ways of knowing. The moral dimension of teacher education is considered as a backdrop for understanding how teachers come to know.

## 1. Introduction

The notion of teacher knowledge is being recognized as an increasingly complex phenomenon compared with the simplicity and naiveté that characterized our notions in past decades. Formerly, our conceptions of teacher knowledge consisted primarily of understanding what teachers knew about mathematics. Reform in mathematics teacher education consisted primarily of providing teachers a better background in mathematics which usually translated into teachers taking more courses in mathematics. Studies by Begle (1968) and Eisenberg (1977) raised our consciousness that effective teaching involves much more than a teacher being mathematically competent. Shulman's (1986) notion of pedagogical content knowledge made explicit what concerned many mathematics educators, namely, that mathematical knowledge alone does not translate into better teaching. Bromme's (1994) transformation of Shulman's conceptions of pedagogical knowledge put a mathematical face on Shulman's more general notion as Bromme focused on the nature of school mathematics, philosophies of mathematics, and other factors that we recognize today as influencing our conception of the mathematics being taught in schools. Lappan and Theule-Lubienski (1994) concluded that,

> Teachers need knowledge of at least three kinds to have a chance to be effective in choosing worthwhile tasks, orchestrating discourse, creating an environment for learning, and

*Educational Studies in Mathematics* **38:** 163–187, 1999.

analyzing their teaching and student learning: knowledge of mathematics, knowledge of students, and knowledge of the pedagogy of mathematics. (p. 253)

Even and Tirosh (1995) see teachers' mathematical knowledge and knowledge about students as mediators in how they teach mathematics.

In this article I will address issues related to teachers' knowing mathematics and knowing about the teaching and learning of mathematics. Some of the evidence I will cite is based on data gathered and analyzed as part of the 'Research and Development Initiatives Applied to Teacher Education' (RADIATE) project.[1] One of RADIATE's goals was to understand the meanings that prospective secondary mathematics teachers (mostly aged 21–23) developed about mathematics and teaching as they progressed through their approximately 15 month instructional program in mathematics education. Our study of the RADIATE teachers included the collection and analysis of survey data, 9 interviews conducted over the 15 months, and observations of the prospective teachers as they participated in university courses and in field activities including student teaching. More detailed analyses and results can be found in Cooney, Wilson, Albright, and Chauvot (1998).

## 2. Examining teachers' knowing

It is only in the past several decades that the field of mathematics education has come to appreciate that the enhancement of what teachers know extends far beyond the domain of mathematics per se. This complicates the process of understanding how teachers come to know mathematics and the teaching of mathematics. I begin my analysis by examining what secondary teachers know and believe about mathematics and the teaching of mathematics.

### 2.1. *Examining what teachers know and believe*

Much of the research on teachers' knowledge leads us in the direction of a deficiency model in that it points out what teachers (often elementary teachers) lack as mathematical knowledge. (See, for example, Wheeler and Feghali (1983) and Graeber, Tirosh and Glover (1986).) If one is trying to understand the status of what teachers know and to build a case for increasing teachers' knowledge of mathematics, this approach has merit. Weiss, Boyd and Hessling (1990) point out that inservice programs can and do increase teachers' mathematical knowledge, thus eliminating, assumedly, part of a perceived deficiency. We have less information about what secondary teachers know about mathematics, although we are accumulating considerable evidence about what they believe (see Thompson, 1992). With respect to what mathematics teachers know, I remember vividly my observation in the early 1970s of a young, bright preservice teacher teaching middle school students about the identity element for addition of whole numbers and rational numbers. She made the point that zero is the identity element for addition of whole numbers, basing her claim on examples involving established addition facts. What caught me off guard was the claim that 0/0 was the identity for the rational numbers since 0/0 + 2/3 = 2/3! I might have dismissed this as a simple mistake had it not been for the fact that she illustrated the principle several times making other errors in the process. My shock was heightened by the fact that this young teacher, who had obviously planned carefully for the lesson, had received top grades in all of her upper level mathematics courses (including abstract algebra); indeed, she graduated with honors from the University. Simply put, she was an outstanding student. How can this come to be? The answer to this question is long and involved, but the short of it is that often preservice teachers have a poor understanding of school mathematics – having last studied it as teenagers with all of the immaturity that implies.

Cooney et al., (1998) provide extensive documentation that preservice teachers lack fundamental understandings of school mathematics despite their success in studying advanced university level mathematics. For example, preservice RADIATE teachers in a survey conducted as they entered their mathematics teacher education program, were asked 'What is a function?' Responses included the following.

A function is a formula that can have items inserted.
Is an algebraic equation of a line in a plane.
A function is an equation with variables and can be plotted on an x, and y graph to see how the graph curves, turns, or doesn't exist at a particular value.

Their notion of function was closely related to the concept of equation and revealed a strong computational orientation. I might add that this equation-oriented notion of function also was evident in a survey conducted with 200 experienced secondary teachers who were asked to create an item that could reveal students' deep and thorough understanding of mathematical functions. Many teachers generated an item that involved solving an equation, made difficult because of parentheses or the presence of fractions, but equation solving nonetheless (Cooney, 1992).

The initial RADIATE survey also revealed that the preservice teachers experienced difficulty in recognizing the graphs of exponential or logarithmic functions (Cooney et al., 1998). When asked why it is important to study functions in high school, the respondents were mostly of a single mind, viz., students will encounter functions in their later study of mathematics so it is important that they get the basics, a sort of circular justification for teaching functions. They had difficulty generating mathematical applications other than the usual textbook problems involving growth and decay although they consistently attested to the importance of mathematics for solving real world problems. What specific applications were given usually involved computations involving sales tax or discounts. Teachers were comfortable solving equations but less so when asked to identify and analyze different kinds of graphs. Although the survey responses were collected during the first class of the first course in mathematics education, and the teachers' knowledge improved throughout the courses, results such as those described do expose the need for teacher education to address making connections within the mathematics the teachers would eventually be teaching. Even with changes evolving, it seems fair to say that the RADIATE teachers saw mathematics as existing outside the human domain – a sort of Platonic view of mathematics. It was common for the teachers to align the doing of mathematics with getting right answers, although some teachers emphasized multiple solution methods for solving problems. One of the teachers, Ken, believed that mathematics consisted of 'certain undeniable truths or laws' and that mathematics may be discovered but could not be created because it already exists. Later in his program Ken allowed a more pluralistic view of mathematics consisting of problem solving and reasoning (Eggleton, 1995). What seems apparent is that preservice teachers do not enter their teacher education programs with a solid grounding in school mathematics or in how mathematics can be applied to real world situations although some teachers emphasized a pluristic view of mathematics and acknowledged the importance of intuition, curiosity, and reasoning in doing mathematics. Even though some teachers' beliefs about mathematics changed over the course of the program, we should caution ourselves not to take for granted that preservice teachers have the depth of mathematical understanding necessary to sustain reform in mathematics education based on their study of collegiate level mathematics alone.

Although evidence is scant, it appears that even experienced secondary teachers are limited in translating their mathematical knowledge into tasks that require a deep and thorough understanding of mathematics. Cooney, Badger and Wilson (1993) found that over one-half of the teachers surveyed produced procedurally difficult but conceptually simple tasks when asked to generate an item that required a deep and thorough understanding of mathematics on the part of the student. For example, an item that teachers claimed would test a minimal understanding of fractions would be to find the sum 3/8 + 1/4 whereas an item that tested a deep and thorough understanding would be to find the sum 4 1/5 + 3 2/3. Similarly, Senk, Beckmann and Thompson (1997) found that about 68% of teachers' tests focus on lower level outcomes and that only about 5% of the items require any depth of thinking. Further, they found that 'Virtually no teachers used open-ended items on tests' (p. 202) a finding consistent with Cooney et al. (1993) who reported that teachers felt uncomfortable in answering and unlikely to use open-ended items with their students. The teachers who participated in these studies may have had strong mathematical backgrounds but even if they did, their translation of that knowledge into meaningful tasks for students is suspect.

Research suggests that teachers' beliefs about mathematics are often limited and perhaps dualistic in the sense of having a right/wrong orientation with mostly single procedures to arrive at the correct answer (see Thompson, 1992). Cooney et al. (1998) found that a consistent theme among the RADIATE teachers was their equating of good teaching with good telling, meaning that kids should understand mathematics step-by-step and should not be confused. As one teacher put it, the responsibility of the teacher is to present clear and concise step-by-step mathematics; the responsibility of the student is to take good notes and do the assigned problems. A second theme that emerged was that of caring. The teachers wanted to be fair and wanted to motivate their students so that they could see the fun in mathematics as they saw it. For some of the teachers, caring meant enabling students to master basic skills. One teacher professed the importance of students making sense through cooperative learning efforts, but, in fact, his classroom teaching was dominated by teacher-telling. His notion of having a comfortable classroom consisted of making tests a matter of mimicking what had been previously demonstrated. Although it is laudable that teachers feel a responsibility to give their students clear directions on how to do mathematics and that they care deeply about their students' learning, this reductionist orientation represents a significant obstacle to most reform movements in mathematics education. In particular, it leads to an isolationist perspective in which new ideas are suspended in favor of the more familiar. Tellling and caring are not the grist that supports the development of a reflective practitioner as conceived by Schön (1983) and others.

It seems clear that the RADIATE teachers learned a considerable amount of mathematics and became more enlightened about the nature of mathematics and how it could be taught as they progressed through the program. In some sense, we were filling a perceived void. But the more relevant issue may be the way that knowledge is held. Filling voids, in isolation from other considerations, will not provide the kind of knowledge base that enables teachers to experience a paradigm shift away from telling and toward a more constructivist oriented approach to teaching. To promote accommodation and not just assimilation or perhaps even rejection, it is important to address the 'casing' in which mathematics is learned, the point of the next section.

### 2.2. *The importance of casing*

It can be argued on both empirical and philosophical grounds that what teachers learn is framed in the context in which that knowledge is acquired. In some sense this seems obvious, for how else can we learn but in a context. But a deeper analysis of teachers' learning should give us some pause about the medium through which teachers acquire knowledge. Preservice teachers come to our teacher education programs with notions about what constitutes teaching. Not surprisingly, their view of the teaching of mathematics is, more or less, consistent with the way they experienced learning mathematics. Most often, this consists of a telling mode of teaching, particularly at the secondary school level. It would be very unusual to encounter a preservice teacher who professes and demonstrates a propensity to teach in what Steffe and D'Ambrosio (1995) refer to as a constructivist mode of teaching. This is not to damn the teachers but rather to say that their learning experiences often counter current reform efforts in mathematics education. Further, novice teachers usually wish to rule with love and sweet reason, thinking that their students will see mathematics much as they see it. But often their idealism and their students' expectations are in conflict. Nearly 30 years ago, Ryan (1970) provided the following analysis.

> In the first year of teaching, then, we witness the sad counterpoise of two sets of attitudes on how the teacher should act. The students are looking for strong personalities and leadership. The beginning teacher, however, seeks a more gentle leadership style. For some few teachers, this works. For the legions, it fails. (p. 181)

Ryan's comment is primarily referring to leadership in the classroom and the teacher's need for stability and order in that classroom. Ask student teachers what concerns them most and it is likley that concern will dwell on some aspect of classroom management. It is not surprising that, when asked the following question to stimulate their thinking about the teaching of mathematics,

> If you could be an animal when teaching, what animal would you be?

two popular answers by preservice teachers are the elephant and the lion justified by the belief that these two animals are strong, powerful and beyond intimidation.

Eventually, the beginning teacher sorts out the kind of leadership that is needed in his/her classroom. But herein lies a problem, namely, the confusion between being the social authority in the classroom and the legitimizer of truth. These two notions of authority become easily blurred. If one controls the intellectual substance of what gets taught and learned, then it becomes much easier to control the social dynamics as well. If the teacher defines his role as that of effective telling, then it likely follows that he sees the students' responsibility as that of listening. Teaching for problem solving is risky business because it invites the unpredictable and raises the question as to how many perturbable events a typical teacher can accommodate without fear of loosing control of the class. There is an inherently strong conservative force that serves as a backdrop for what and how teachers come to know. Left unattended it seems quite reasonable that the inertia resulting from this force will reveal that the teaching of tomorrow's mathematics will appear eerily like that of yesteryear. We may substitute different mathematics and it may be enhanced by technology, but the basic delivery system will be essentially the same.

When one of my undergraduate advisees indicates that she has an excellent mathematics professor, I often ask what makes the professor so good. The answers are usually the same and focus on the following teacher characteristics: clarity, pace of instruction, and availability for answering questions. These are laudable characteristics to be sure. Further, they are often empirically associated with effective teaching (see Cooney, 1980). They are, however, not the substance from which reform is derived, for they embrace an environment of telling. Teaching in the schools would be much simpler if teachers were asked only to deliver instruction based on these three principles. Perhaps it might be easier, but reform it is not.

In their desire to be a 'brilliant' mathematician in the eyes of their students, the teachers strive to adopt these same elements in their own teaching. They honor the professors they have had and, in many ways, want to emulate them. This conservative perspective is often buttressed by the classroom teachers with whom they work in the field as part of their teacher education program. Jones and Vesilind (1996) found that the single most influencing factor in shaping preservice middle school teachers' beliefs was that of student teaching, with university courses being a very distant second. For example, the concept of flexibility was initially associated with preparation for teaching. But this construct evolved into attention to the unpredictable and later in student teaching to differentiating instruction to meet individual students' needs. There is much that is positive about this although other studies have indicated that the impact of student teaching on teachers' beliefs is often a very confining one (Cooney, 1980). What seems undeniable is that what got learned was shaped to a great extent by the context in which that learning took place, most notably, experiences in the field coupled with teachers' own learning experiences.

Teacher education is consequently in the unenviable position of needing to help teachers unravel their notions about teaching and rebuild them in a rational way. Enter the notion of reflection. But the orientation toward telling with clarity and the overwhelming propensity to be a caring teacher puts at risk ideas that may appear to contradict these characteristics. How can it be, for example, that caring can be translated into causing students to experience stress in solving problems? Clarity is a very seductive characteristic in teaching for it satisfies both teachers and students – not to mention parents and administrators. Never mind that the resolution of conflict constitutes an important human activity. Clarity calls for a classroom in which learning is predictable and, most likely, atomistic in nature. This is not to say that clarity is the enemy of good teaching, but it is to suggest that clarity without rationale can lead to a static form of learning. The case of Sue (Cooney, 1994a) illustrates that it is not easy for an experienced and good teacher, certified so by students, parents, and administrators, to admit that a constructivist orientation toward learning has merit and requires a shift away from a teacher-centered classroom. For Sue, the shift required the dismantling of much of what she had previously valued in favor of constructing a different philosophy of teaching – one that could be both more productive and more humanistic. Sue was a remarkably talented teacher who could endure and prosper from the unpacking and repacking. But Sue is not your typical teacher and consequently we are faced with a considerable challenge in teacher education.

## 3. Conceptualizing Teachers' Belief Structures

As Even and Tirosh (1995) remind us, categories of knowledge and beliefs often get blurred when we study teachers in their working environment. As Pajares (1992) has argued, if we are interested in understanding human behavior we must recognize that 'knowledge and beliefs are inextricably intertwined, but the potent affective, evaluative, and episodic nature of beliefs makes them a filter through which new phenomena are interpreted' (p. 325). Teachers operate in contexts, their knowledge framed and shaped by experiences many of which happen long before they formally enter the world of mathematics education. In the classroom, what the teacher knows is fused with her sense of purpose as a teacher of mathematics, her philosophy of teaching and learning, and her sense of responsibility given the community in which she teaches.

If a teacher's knowledge base is to support her becoming an adaptive being (Cooney, 1994b), that knowledge must be of the form that Belenky, Clinchy, Goldberger and Tarule (1986) construe as constructed knowledge in which the voices of others are integrated into a consistent whole or in terms of what Perry (1970) calls *relativistic thinking*. It is from this integrated perspective that one has the capacity to acknowledge context as a mediating factor in conceptualizing and acting out a course of action in the classroom. This action involves reflection, the vehicle for becoming an adaptive agent. Schön's (1983) notion of reflecting-on-action and reflecting-in-action characterize the kind of reflective thinking that leads one to become an adaptive teacher who attends to context. To von Glasersfeld (1991), reflection is the ability of an individual to 'step out of the stream of directed experience, to re-present a chunk of it, and to look at it as though it were direct experience, while remaining aware of the fact that it is not' (p. 47). Above all else, the ability to be reflective in the sense of which Schön and von Glasersfeld speak is paramount to a teacher's professional development. Consequently, I submit that whatever lens we use to describe teachers' knowledge, that lens must account for the way in which knowledge is held and the ability of the teacher to use that knowledge in a reflective, adaptive way.

Green's (1971) metaphorical analysis of beliefs provides one vehicle for conceptualizing the structure of beliefs. His analysis consists of thinking of beliefs being held in a spatial sense (psychologically central versus derivative) and in a quasi logical sense (primary versus derivative). Green's analysis is grounded in the notion of how we come to believe. Do we believe A because we believe B and we believe B because someone told us this is the case? Or do we believe A on the basis of either direct evidence or on the basis of belief B which is grounded in either empirical or rational evidence? The answers to these questions have much to do with Green's differentiation between indoctrination and teaching. In the former, one comes to know because an authority (teacher, textbook, professor) has said it is so. The latter is based on allowing the individual to come to know based on his/her own experiences perhaps guided but not imposed by the teacher. Teaching methodologies based on the former emphasize telling and memorizing; teaching methodologies grounded in the latter emphasize the processes of doing mathematics, as suggested in recent reform documents, and reflect the notion of mathematics as a fallible science.

Now we can see the rub of the matter. If we desire that teachers become reflective practitioners in which mathematics is seen as a creation of knowledge rooted in rationality, then it seems obvious that the teacher's knowledge must be of a form that supports such a perspective. Complicated as it is, we should at least recognize the role that context and reflection play in allowing the fluidity and flexibility of knowledge that permits reform. The issue is not just what the teacher knows but the means by which that knowledge is acquired. That is, the medium and the message are inseparable. The process of learning is fundamentally connected to how beliefs are structured, whether the beliefs are rooted in rationality or are the consequent of telling.

Cooney, Shealy and Arvold (1998) have developed a characterization of teachers based on their case studies of preservice teachers. Their characterizations are rooted in the work cited above combined with the contributions by Dewey (1933), Schön (1983), and von Glasersfeld (1991) on the notion of reflection. The four characterizations are *isolationist*, *naive idealist*, *naive connectionist*, and *reflective connectionist*. An isolationist tends to have beliefs structured in such a way that beliefs remain separated or clustered away from others. Accommodation is not a theme that characterizes an isolationist. For whatever reason, the isolationist tends to reject the beliefs of others at least as they pertain to his/her own situation. More importantly, this rejection is rooted in a circular array of beliefs impervious to empirical evidence. In short, it characterizes a person who 'knows' the right way to teach when he/she enters a teacher education program and sees little value in what any such program has to offer other than beliefs consistent with his/her own. The naive idealist tends to be a received knower in that, unlike the isolationist, he/she absorbs what others believe to be the case but often without analysis of what he/she believes. This position is consistent with Belenky, Clinchy, Goldberger and Tarule's (1986) 'received knowers' who rely on others to provide the evidence and substance of what they know. Shealy's (1994/1995) Nancy exemplifies a naive idealist. A prominent characteristic of Nancy was her assumption of consensuality. She defined much of her knowledge about teaching in terms of others' voices – classmates, professors, significant others. She did not reject ideas as does the isolationist. Rather, she uncritically accepted them.

In contrast to the naive idealist, the two connectionist positions emphasize reflection and attention to the beliefs of others as compared to one's own (Cooney et al., 1998). The naive connectionist fails to resolve conflict or differences in beliefs whereas the reflective connectionist resolves conflict through reflective thinking. The reflective connectionist position was exemplified by the case of Greg, a preservice teacher, who began his teacher education program with a negative view of technology but later incorporated his strong support for the use of technology into his core belief of wanting to prepare students for life (see Shealy, 1994/1995). Perhaps this would have happened anyway, but the continual opportunities afforded

Greg for reflective activity in a mathematical-pedagogical context may have accentuated his modification of beliefs. Because Greg readily made connections between different ideas and provided consistent evidence of being a relativistic thinker, we conceptualized the reflective connectionist position. In general, a reflective connectionist integrates voices (including one's own) and analyzes differences consistent with Perry's (1970) notion of relativistic thinking. The connectionist positions are contrary to teaching as telling as a primary instructional mode because of their emphasis on context. Although the notion of caring could be consistent with a connectionist perspective, it would not be if caring precludes the discomfort of encountering challenging ideas.

We can see, then, the struggle that defines teacher education, viz., transforming the preservice teachers' notions of telling and caring into ones that encourage attention to context and reflection. We should recognize that the counterpoint of telling should not be 'never telling,' a perspective that is not only irresponsible but also belies classroom reality. Context determines when telling is appropriate and when it might impede a student's mathematical development. Similarly, preservice teachers' notion of caring needs to be transformed into a construct that attends to the student's intellectual being and not just his/her comfort level. The connectionist appreciates the distinction between comfort and intellectual progress and how one does not have to be subserviant to the other.

The generality of these four positions is yet to be determined, but the theoretical perspective underlying them seems promising as a way of conceptualizing the structure of teachers' beliefs. Consider the cases of Harriet and Kyle (Cooney and Wilson, 1995), both RADIATE teachers. Harriet tended not to be a very reflective individual, at least as she presented herself in our data. We detected five factors that seemed to influence her beliefs. In order of their perceived importance to Harriet, they were: her mother, field experiences within the teacher education program, technology, conversations with peers, and the teacher education program as a whole. One might argue that the teacher education program was a more significant influencing factor since it provided the context in which field experiences and conversations with peers occurred. But if by teacher education program we mean the ideas and values that were continually emphasized throughout the program, then it seems clear that the teacher education program was not a very important factor. Consider, for example, that in her last interview in which she was to have read transcripts from the previous eight interviews and identify what she said in those interviews that she deemed important, she identified only four statements (most teachers identifed 20 or more statements). Three of those statements were related to ideas from her mother's way of teaching mathematics, ideas expressed in the first interview at the beginning of the program! In some sense Harriet was an isolationist in that she held at bay most of the ideas she encountered during her teacher education program. Yet, she did feel that technology had an important role in the teaching of mathematics and she did acknowledge that it was meaningful to share ideas about teaching mathematics with her peers. In somewhat of a contrast, Kyle was more reflective and recognized inconsistencies in his own beliefs about teaching mathematics. For example, he felt that it was quite important to help students learn how to solve real world problems. He solved real world problems himself and enjoyed that aspect of doing mathematics. Yet, he was critical of his collegiate calculus teacher, who spent a considerable amount of time solving real world problems, for not helping him learn the basics. His reflective nature suggested he was some sort of a connectionist albeit of what kind is not clear.

It is important to understand that Harriet and Kyle were indistinguishable in terms of their credentials in mathematics. They both had essentially the same mathematics courses with more or less the same grades. Their overall academic performances at the university were very similar. What separated them was the nature of their reflective thinking and the approach they took to new ideas. That is, the way they held their knowledge was more of a distinguishing characteristic than was the knowledge itself. Because he tended to be more reflective, Kyle's beliefs were permeable; Harriet's were not. Their mode of thinking continued into their first year of teaching. Harriet was uncritically accepting of her teaching performance which had a heavy emphasis on procedural knowledge and was decidedly teacher centered. Kyle, on the other hand, was critically unaccepting of his teaching performance. He recognized that his teaching was teacher centered with little emphasis on the kind of mathematics he valued. He was very unhappy with this circumstance yet, largely because of disciplinary problems, felt unable to change his teaching style. At one point, he was seriously considering dropping out of the profession because his teaching was not reflecting his vision of what a classroom should be like.

These and other cases provide strong evidence that the structure of one's beliefs is an important factor in determining what gets taught and how it gets taught. Consequently, it ought to be an explicit focus of teacher education programs. In the next section, I will discuss at least one attempt to influence the structure of teachers' beliefs so as to promote the development of a more reflective and adaptive teacher.

## 4. The Notion of Integrating Content and Pedagogy

I have argued that preservice secondary mathematics teachers exhibit a fragile view of school mathematics. Often their lesson planning consists of working the examples they plan to use and making sure they can do the assigned problems. Missing is the notion of helping students make mathematical connections between yesterday's mathematics and tomorrow's. They are good students and have survived, if not mastered, significant courses in mathematics. They are eager to teach mathematics in a way that fits their definition of good teaching. They care. But they also lack not only a deeper knowledge of school mathematics but the kind of pedagogical content knowledge of which Shulman (1986) speaks. With the recognition that the medium and the message are inseparable in terms of what gets learned, I submit that the integration of content and pedagogy is of primary importance in directing the creation of materials to help alleviate the curricular problem we have in mathematics teacher education and to influence teachers' ways of knowing so as to promote a more reflective orientation toward teaching. Our cries for a constructivist-oriented teaching style often fall on deaf ears given the immediacy of the problems teachers face in the classroom. As one young preservice wrote in her journal, 'It is difficult to listen to what the students are saying when I am worried about what to say next.' This suggests that attention to mathematics in the absence of pedagogical considerations will likely fall short of enabling teachers to acquire the kind of knowledge that supports listening to students.

Where do we plant the seeds for realizing a more student-centered classroom? I suggest that it begins with the teacher reflecting on what mathematics means to her and how she envisions the teaching of mathematics. I have used the following reflective situations as course 'openers.' They serve as contexts for teachers to begin thinking reflectively. The two items also serve as a context for subsequent interviews in which the RADIATE researchers explored teachers' beliefs about

---

Exploration 1

Find a willing friend and ask him or her the following question: If you could think of something that is as different from mathematics as possible, what would it be? Explore why he or she picked whatever was picked. Write a one-page report on what you think your friend's view of mathematics is. (Cooney, Brown, Dossey, Schrage and Wittmann, 1996, p. 5)

---

Reflective problem 4

Consider analogies with the following possibilities and decide which one(s) best fit your notion of what it means to be a mathematics teacher. Provide a rationale as to why you made the selection that you did.

| | | |
|---|---|---|
| newscaster | orchestra conductor | physician |
| missionary | gardener | engineer |
| social worker | entertainer | coach |

How does your selection and rationale compare with those of your classmates?

(Cooney, Brown, Dossey, Schrage and Wittmann, 1996, p. 14)

mathematics and its teaching. The question in Exploration 1 is sometimes posed directly to teachers when the question is part of an interview protocol. The importance of the questions lie not in the actual responses they evoke but in the rationale provided for the responses. We have found that these and other similar questions provide a rich research site for exploring teachers' beliefs.

If we take seriously the notion of mathematics as the science of studying patterns, then mathematics so defined should permeate any level of study from elementary school through graduate school. The connections we expect teachers to make and promote in their classrooms should be an integral part of *their* mathematical experiences. But the context in which the mathematics is learned should have a pedagogical flavor as well. Cooney, Brown, Dossey, Schrage and Wittmann (1996) have written materials that promote reflection (see Reflective Problem 4) and provide definition for integrating content and pedagogy. We have adopted the approach of embedding mathematical problems in pedagogical contexts so that there are essentially two problems to solve: the mathematical problem and the pedagogical problem. For example, in one episode Ms. Lopez presents the following 'biggest box problem' to her students.

The biggest box problem

What size square cut from the corners of the original square maximizes the volume of the figure formed by folding the figure into a box without a top?

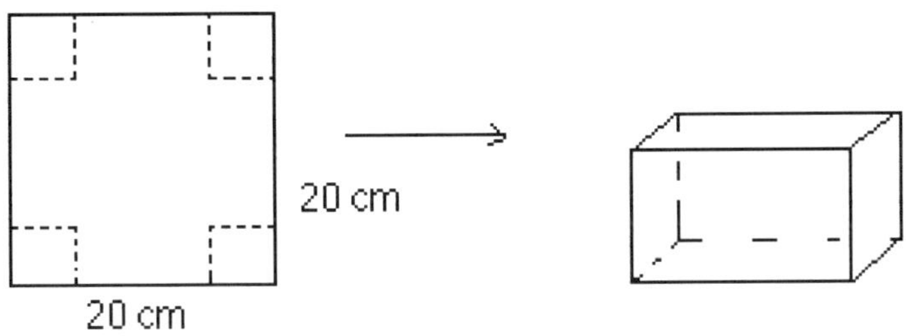


It is a problem familiar to most calculus students. But our approach is different. The classroom dialogue shows Ms. Lopez's students solving the problem using spreadsheets and graphing technologies. It is a wonderful lesson with students actively contributing mathematical ideas. But then the question arises as to what the exact solution is. Ms. Lopez has forgotten how to solve the resulting cubic equation by algebraic means and fumbles around trying to uncover relevant mathematics. Predictably, the lesson begins to unravel. Students become impatient when they see that Ms. Lopez is losing control of the intellectual ship. Now we have a pedagogical problem as well as a mathematical one. We could suggest that had Ms. Lopez remembered how to solve the problem algebraically, the smooth lesson would likely have continued without a hitch. But this begs the question of what a teacher should do when someone asks a question for which an immediate answer is not known. What interests the teachers who read the Lopez story is not only figuring out the problem without using calculus but also considering how and in what context they could use the box problem when they begin teaching.

The mathematical problem can be extended to include finding the dimensions of the biggest box when the surface area of the box remains 400 square centimeters. But at some risk. We have found that the preservice teachers' reactions to this extension is mixed. Some relish the extension because it represents challenging mathematics. On the other hand, some complain that the extension belabors a problem that has little to do with the teaching of mathematics. Indeed, what we have is a microcosm of the secondary school classroom in which a teacher promotes problem solving and problem extension, a point discussed during the course and explored as part of our research program.

Another problem that we have used with our teachers is the 'pentagon problem.' It raises questions about similarity along with other mathematical issues. We use the problem to provide a context in which we can explore how and why such problems could be used with secondary school students. A typical question is, "What mathematics would you expect secondary students to learn and how would you assess what they learned?" Further, we ask them to create a

The pentagon problem

What is the largest possible pentagon with the same shape as the one shown below that could be drawn on a regular sheet of typing paper? Explain and justify your solution method.

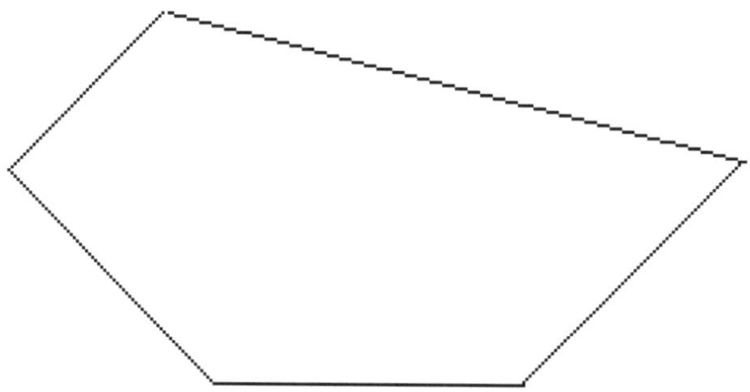

context in which the problem could be framed in a real world situation. (Drawing a model of a lot for building a house is one such frame.) The problem also reveals misconceptions about similarity as some students maintain that figures whose sides are constructed parallel to the given sides and a specified distance (say 3 cm) from the sides results in a figure similar to the given figure. How does one determine whether this conjecture is true or false?

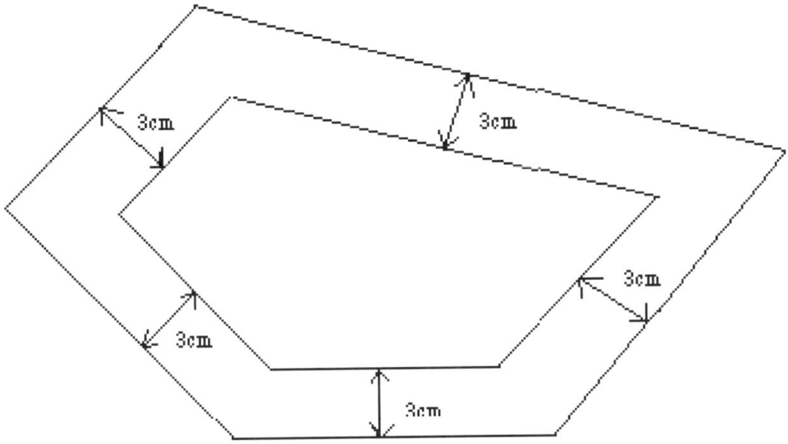


Is the larger pentagon similar to the smaller pentagon? Why or why not? Subsequently we consider how the problem can be varied to accommodate various teaching situations. Some possible alternatives include the following.

a. What is the largest square that could be drawn on a regular sheet of typing paper? (Sides of square parallel to sides of paper.)
b. Given the dimensions of a rectangle (say 3 cm by 6 cm), what is the largest similar rectangle that could be drawn on the typing paper. (Sides of rectangle parallel to sides of paper.)
c. Problems a and b could be repeated but the square and the rectangle could be drawn so that their sides are oblique to the sides of the paper.

The same pedagogical questions remain as to the context and anticipated outcomes, but the focus of the discussion now changes to how problems can be modified to accommodate different classroom situations. It provides a context for considering how mathematical/pedagogical alternatives can be generated.

In another kind of learning environment I have presented teachers with a variety of construction problems (e.g., Construct a square given a segment that represents the sum of its diagonal and side) and asked them to solve the problems in cooperative learning groups of size four. Each group's assessment is based on a random selection of the teacher who will be assessed and the problem to be solved. The teachers have to rely on each other, if not to solve the problems initially, then at least to understand the construction and its proof for the assessment. Typically, each group assigns to its members different responsibilities, with some solving one set of problems, others solving a different set of problems. It matters less who solves which problems and more that each teacher learns how to solve every problem, some by solving the problem initially and others by learning from their peers. This context provides real meaning to cooperative learning groups for it is in the best interest of the entire group to make sure everyone can solve the problems.

In all of these situations, I have tried to engage teachers in an interesting and challenging mathematical situation which is embedded in a pedagogical context. Preservice teachers react positively and demonstrate an increased competence in doing secondary school mathematics beyond what is typically the case as described earlier. Perhaps the greater issue is whether we have helped them attend to the potential context in which they will be teaching mathematics and to develop a more relativistic view of mathematics. Our evidence suggests these teachers can become connectionists and perhaps even reflective connectionists. Nevertheless, the conservative nature of the student-teaching experience and the potential isolation during the first year of teaching cast a long shadow over potential reform efforts. What seems critical is that teachers see something problematic about the doing, teaching, and learning of mathematics. What gets tricky is that their own drive for simplicity and certainty can work against efforts to reform. Life in the classroom is much easier when events unfold as expected, as the episode with Ms. Lopez demonstrates. For a first year teacher, predictability is a friend, albeit it can also be a seed for classroom sterility. The means by which we come to know is perceived to be more friendly when the road is less bumpy. I recall interviewing a teacher who indicated a strong preference for students learning how to solve problems in his class. When asked how he expected to do this, his response was 'In as clear and efficient manner as possible.' One can only wonder what his notions of problem solving were and what his students perceived problem solving to be.

## 5. The Moral Dimension of Knowing

At first glance, Green's (1971) distinction between indoctrination and teaching might seem inappropriate when applied to the teaching of mathematics. Indoctrination seems like a strange way to describe the teaching of mathematics. Yet if indoctrination refers to the process of knowing in which what gets learned is determined and defined by an authority, then it seems quite applicable to the teaching of mathematics. What does it mean to know the Pythagorean Theorem? Although the answer entails many dimensions, most would agree that citing the equation is $a^2 + b^2 = c^2$ is hardly adequate evidence for claiming to know the theorem. It should be kept in mind that mathematics presented from a dualistic perspective does little to engage students in the kind of mathematical thinking that leads to generative knowledge. For a democratically-oriented society, the power to reason and to make judgments is critical and should be a central part of the educational system's mission.

Ball and Wilson (1996) argue that when it comes to teaching, the intellectual and the moral are inseparable. They point out that the entrance to the road of ideas is not totally a function of the ideas themselves but rather the means by which that entrance is gained. As they put it,

> Intellectual honesty implies engaging students in the conjecturing, investigating, and argument that is characteristic of a field. But responsibility to students means grappling with the consequences of students reaching conclusions that their next teacher will see as wrong. How should teachers reconcile an emphasis on reasoning with a concern for particular ideas? (p. 182).

The issue, as pointed out by Ball and Wilson (1996), is much deeper than considering a student's proposition that seems reasonable on the one hand yet is false. How can we engender in students ways of validating claims that appear to be reasonable? Ball and Wilson suggest that 'Although ignoring students' nonstandard inventions may be to deny them respect, withholding students' access to other perspectives is no simple alternative' (p. 186).

There is, of course, a symmetrical issue in teacher education. Most of the argument provided in this article has focused on moving teachers toward a more reflective and adaptive perspective, a perspective that may not be immediately acceptable to the young teacher. An interesting twist in all of this is that not infrequently preservice teachers see themselves as answering a call to become a teacher – a moral imperative to teach. Helms (1989) observed that one of his teachers, Al, believed that it was important to get to know his students' interests beyond the mathematics classroom, a position he claimed was rooted in his religious beliefs as he saw students as God's creatures. His continual emphasis on the importance of nurturing students stemmed from his belief that the role of the church was to provide a context of support in the form of a spiritual family. Interestingly, Al also believed that God created mathematics in a pure form devoid of human endeavor. As Al put it, 'Math itself is not involved with human frailties' (p. 154). Tabatha, another preservice teacher, decided to change majors and become a mathematics teacher because there was a 'very small voice in the back of my mind' a voice related to God's way of speaking to her (Helms, 1989). Both of these teachers revealed linkages between their beliefs about mathematics and its teaching and their religious beliefs. Although the connections noted by Helms were more explicit than those expressed by the RADIATE teachers, we too noted that several of the RA-

DIATE teachers communicated a strong sense of morality that shaped their views about their responsibilities as teachers. Interestingly, these teachers denied that morality would influence the means by which they would teach mathematics since mathematics itself was an amoral subject. They would not impose their moral directions on their students when it came to the more sensitive and explicit issues addressed in the media.

What is there about the teaching of mathematics that is moral if the subject itself is perceived (by the teachers) to be amoral? What is there about any subject that makes it moral, immoral, or amoral? For most secondary school subjects, the answer lies not in the content itself but rather in how one comes to know that content. It has to do with evidence and beliefs. Green (1971) provides the following analysis with respect to connecting evidence and beliefs.

> Given any two beliefs, A and B, a person may believe A because its truth is supported by B. In this case, he accepts a certain belief because there is evidence in support of it. He accepts it on the basis of that evidence. But it can also happen the other way around. A person who believes A in the first place, might believe B because he thinks it supports a belief that he already holds. In this case, he does not accept a belief because it is supported by evidence; instead he tends to accept a certain belief because he thinks it will lend support to some other belief he already accepts. In other words, *a person may hold a belief because it is supported by the evidence, or he may accept the evidence because it happens to support a belief he already holds* (Author's emphasis) (p. 49).

An internally consistent set of beliefs, circularity justified, can lead to what Rokeach (1960) refers to as a closed mind. Dogmatism, the consequence of having a closed mind, is built on a sequence of beliefs held nonevidentially. It is counter to rational thinking and results in impermeable beliefs. Although it does not follow that a Platonic view of mathematics leads to teaching mathematics from a nonevidential perspective, it is not a long leap from the Platonic view to the position that the teacher's role consists of informing students about mathematical truth in a nonevidential way. Such a view of mathematics and its teaching is contrary to most of the reform movements and to the kind of teaching addressed by Clarke (1997) or by Lloyd and Wilson (1998).

A few years ago I was involved in analyzing the mathematics instruction in an elementary school. I observed every teacher's teaching of mathematics in grades K-3 at least once. On the one hand, I had a very good feeling about what I was observing. The teachers and the students appeared interested in what was happening and the children enjoyed giving answers for the questions posed by the teachers. The teachers were proud of what they were doing and, in some sense, rightly so. But when I had a moment to review my notes and reflect on my observations, I realized that I had witnessed only two instances over three days of observations in which questions were posed that required something other than immediate recall. There were no instances of students discussing mathematics with each other. In almost every case either the lesson was drill and practice (cleverly done, nonetheless) or else the mathematics was reduced to bits and pieces in which memory could suffice for delivering the correct answer. I felt a sense of moral dilemma. Do I congratulate them on having children excited about 'doing' mathematics? Or should I struggle to weave into my report what I saw as a potential problem for students' later learning of mathematics? Imagine what might happen when the students first encounter a real mathematics problem for which an immediate answer is not forthcoming. I suspect that they will conclude something is wrong–with the teacher, with the curriculum, or with themselves. Talk about a moral dilemma!

What is our moral definition as teacher educators who educate teachers who in turn educate students about mathematics? There can be no mistake that decisions such as what mathematics gets taught, who has access to it, and how does it get taught are, in fact, moral decisions. Perhaps more implicit but just as important is the question of *whose* mathematics gets taught. Wilson and Padron (1994) argue for a culture-inclusive mathematics depicting a mathematics representative of more than the Western mathematics typically taught in our classrooms. We encourage teachers to be reflective and to develop a relativistic view of mathematics and of the teaching of mathematics. Such an orientation seems consistent with democratic ideals and the emerging notion that mathematics is a human endeavor of many colors and origins. The formalism by which mathematics is typically taught is legendary and provides a formidable challenge for teacher education programs. How is it that teacher education can intercede and at least point to other facets and scenarios for the teaching of mathematics? Indeed, what should our teacher education programs be about? Producing caring and mathematically competent teachers? Producing teachers who have the potential to become reflective practitioners? We seem to have less difficulty denying teachers entree into their chosen field if they can not convince us or some agency that they are mathematically competent at least at some level. But what about the isolationist who holds a dualistic view of mathematics and who steadfastly maintains that good teaching is about effective telling? What questions would we consider relevant about that teacher's contributions to a democratic society? Do we have the right to even ask that question as a potential criterion for judging that teacher certifiable as a teacher of mathematics in the public schools?

There are many issues lurking beyond the moral dimension of teacher education. Perhaps the greatest of them all is the challenge we face in enabling teachers to see knowledge acquisition as power so that they can enable their students to acquire that same kind of power. If teachers' ways of knowing are rooted in a cycle of received knowing, then it is predictable that their students' ways of knowing will be received as well. This presents a significant moral dilemma in that the received knower has less intellectual control over decisions that might affect his/her life. To the extent that we think that the goal of teaching is to educate, then we have an obligation to provide our teachers with a similar education.

## 6. A FINAL COMMENT

It is not surprising that preservice teachers' ways of knowing vary as they enter and exit a teacher education program, nor that there is considerable variance among experienced teachers who participate in inservice programs. What seems more to the point is that we develop ways of conceptualizing these differences so as to contribute to our insight and wisdom about how we can encourage reflection and adaptability. I have offered several suggestions as to how we have tried to influence and to stimulate teachers' reflecting thinking about their own beliefs about mathematics and the implications for those beliefs when dealing with both mathematical and pedagogical situations. We have tried to engender a certain element of doubt into the teaching of mathematics without undermining the confidence of the teacher. Reform in the classroom does not grow out of a laissez faire attitude, nor out of programs whose primary intent is to promote conformity to existing conditions often couched in the more technical aspects of teaching. What would benefit teacher education and move it beyond an activity and toward an arena for disciplined inquiry is a framework for conceptualizing teachers' ways of knowing as they progress through their various educational programs. We could then develop programs in which our decisions are informed by more than our intuition, as good (or bad) as that may be. From this perspective, efforts to conceptualize teachers' ways of knowing becomes a moral imperative of the first magnitude.

Ours is a moral profession despite the often perceived amorality of our mother subject. As Brown (1996) has pointed out, logic has its limits, thereby suggesting that we honor connections between mathematics and human experience. Whether or not we embrace such connections, we are shrouded in the moral dimension of teaching and teacher education. Better that we recognize this circumstance rather than further the illusion that mathematics and morality have no common ground. We should be ever mindful that such an illusion leads us down the path of dishonoring the very thing we cherish: the intellectual and moral development of our children's teachers.

## NOTES

1. Project RADIATE (1993–1997) was supported by a grant from the National Science Foundation (NSF) grant # DUE9254475. Views expressed in this article are not necessarily those of NSF. RADIATE was a multisite project involving the University of Georgia, Georgia State University, University of Michigan, University of Rochester, and SUNY at Buffalo. Appreciation is expressed to Patricia Wilson, Co-director of RADIATE at the Georgia site, Maureen Albright, Bridget Arvold, Jennifer Chauvot, Vivian Moody, and Pam Turner for their assistance in analyzing RADIATE data.

## REFERENCES


Ball, D. and Wilson, S.: 1996, 'Integrity in teaching: Recognizing the fusion of the moral and intellectual', *American Educational Research Journal* 33(1), 155–192.

Begle, E.: 1968, 'Curriculum research in mathematics', in H. Klausmeier and G. O'Hearn (eds.), *Research and Development toward the Improvement of Education*, Dembar Educational Research Services, Madison, WI, pp. 44–48.

Belenky, M., Clinchy, B., Goldberger, N. and Tarule, J.: 1986, *Women's Ways of Knowing: The Development of Self, Voice, and Mind,* Basic Books, New York.

Bromme, R.: 1994, 'Beyond subject matter: A psychological topology of teachers' professional knowledge', in R. Biehler, R. Scholz, R. Strässer and B. Winkelmann (eds.), *Didactics of Mathematics as a Scientific Discipline*, Kluwer Academic Publishers, Dordrecht, The Netherlands, pp. 73–88.

Brown, S.: 1996, 'Towards humanistic mathematics education', in A. Bishop, K. Clements, C. Keitel, J. Kilpatrick and C. Laborde (eds.), *International Handbook of Mathematical Education*, Kluwer Academic Publishers, Dordrecht, The Netherlands, pp. 1289–1321.

Clarke, D.: 1997, 'The changing role of the mathematics teacher', *Journal for Research in Mathematics Education* 28(3), 278–306.

Cooney, T.: 1980, 'Research on teaching and teacher education', in R. Shumway (ed.), *Research in Mathematics Education*, National Council of Teachers of Mathematics, Reston, VA, pp. 433–474.

Cooney, T.: 1992, *A Survey of Secondary Teachers' Evaluation Practices in the State of Georgia*, Unpublished manuscript, College of Education, University of Georgia, Athens.

Cooney, T.: 1994a, 'In-service programs in mathematics education', in S. Fitzsimmons and Kerpelman (eds.), *Teacher Enhancement for Elementary and Secondary Science and Mathematics: Status, Issues, and Problems*, Center for Science and Technology Policy Studies, Cambridge, MA, pp. 8.1–8.33.

Cooney, T.: 1994b, 'Teacher education as an exercise in adaptaton', in D. Aichele and A. Coxford (eds.), *Professional Development for Teachers of Mathematics*, National Council of Teachers of Mathematics, Reston, VA, pp. 9–22.

Cooney, T., Badger, E. and Wilson, S.: 1993, 'Teacher education as an exercise in adaptaton', in N. Webb and A. Coxford (eds.), *Assessment in the Mathematics Classroom*, National Council of Teachers of Mathematics, Reston, VA, pp. 239–247.

Cooney, T., Brown, S., Dossey, J., Schrage, G. and Wittmann, E.: 1996, *Mathematics, Pedagogy, and Secondary Teacher Education: Reweaving the Frayed Braid*, Heinemann, Portsmouth.

Cooney, T., Shealy, B. and Arvold, B.: 1998, 'Conceptualizing belief structures of preservice secondary mathematics teachers', *Journal for Research in Mathematics Education* 29(3), 306–333.

Cooney, T. and Wilson, P.: 1995, 'On the notion of secondary preservice teachers' ways of knowing mathematics', in D. Owens, M. Reed and G. Millsaps (eds.), *Proceedings of the Seventeenth Annual Meeting of the North American Chapter of the International Group for the Psychology of Mathematics Education*, ERIC, Columbus, Ohio, pp. 2.91–2.96.

Cooney, T., Wilson, P., Albright, M. and Chauvot, J.: 1998, 'Conceptualizing the professional development of secondary preservice teachers', *Paper presented at the annual meeting of the American Educational Research Association*, April 1998, San Diego, CA.

Dewey, J.: 1933, *How we Think: A Restatement of the Relation of Reflective Thinking to the Educative Process*, D.C. Heath, Boston.

Eisenberg, T.: 1977, 'Begle revisited: Teacher knowledge and student achievement in algebra', *Journal for Research in Mathematics Education* 8, 216–222.

Eggleton, P.: 1995, 'The evolving mathematical philosophy of a preservice mathematics teacher'. Doctoral dissertation, University of Georgia.

Even, R. and Tirosh, D.: 1995, 'Subject-matter knowledge and knowledge about students as sources of teacher presentations of the subject-matter', *Educational Studies in Mathematics* 29, 1–20.

Graeber, G., Tirosh, D. and Glover, R.: 1986, 'Preservice teachers beliefs and performance on measurement and partitive division problems', in G. Lappan and R. Even (eds.), *Proceedings of the Eighth Annual Meeting of the North American Chapter of the International Group for the Psychology of Mathematics Education*, Michigan State University, East Lansing, MI, pp. 262–267.

Green, T.: 1971, *The Activities of Teaching*, McGraw-Hill, New York.

Helms, J.: 1989, 'Preservice secondary mathematics teachers' beliefs about mathematics and the teaching of mathematics: Two case studies', Doctoral dissertation, University of Georgia.

Jones, M. and Vesilind, E.: 1996, 'Putting practice into theory: Changes in the organization of preservice teachers' pedagogical knowledge', *American Educational Research Journal* 33(1), 91–117.

Lappan, G. and Theule-Lubienski, S.: 1994, 'Training teachers or educating professionals? What are the issues and how are they being resolved?' in D. Robitaille, D. Wheeler and C. Kieran (eds.), *Selected lectures from the 7th International Congress on Mathematical Education*, Les Presses de L'Université Laval, Sainte-Foy, pp. 249–262.

Lloyd, G. and Wilson, M.: 1998, 'Supporting innovation: The impact of a teacher's conceptions of functions on his implementation of a reform curriculum', *Journal for Research in Mathematics Education* 29(3), 248–274.

Pajares, M.: 1992, 'Teachers' beliefs and educational research: Cleaning up a messy construct', *Review of Educational Research* 62, 307–332.

Perry, W.: 1970, *Forms of Intellectual and Ethical Development in the College Years*, Holt, Rinehart, and Winston, New York.

Rokeach, M.: 1960, *The Open and Closed Mind*, Basic Books, Inc, New York.

Ryan, K.: 1970, *Don't Smile Until Christmas*, University of Chicago Press, Chicago.

Schön, D.: 1983, *The Reflective Practioner: How Professionals Think in Action*, Basic Books, New York.

Senk, S., Beckmann, C. and Thompson, D.: 1997, 'Assessment and grading in high school mathematics classrooms', *Journal for Research in Mathematics Education* 28, 187–215.

Shealy, B. E.: 1995, 'Conceptualizing the development of two first-year secondary mathematics teachers' beliefs'. Doctoral dissertation, University of Georgia, 1994, *Dissertation Abstracts International, 56–3A,* p. 856.

Shulman, L.: 1986, 'Those who understand: Knowledge growth in teaching', *Educational Researcher* 15, 4–14.

Steffe, L. and D'Ambrosio, B.: 1995, 'Toward a working model of constructivist teaching: A reaction to Simon', *Journal for Research in Mathematics Education* 26, 146–159.

Thompson, A.: 1992, 'Teachers' beliefs and conceptions: A synthesis of the research', in D. Grouws (ed.), *Handbook of Research on Mathematics Teaching and Learning*, MacMillan, New York, pp. 127–146.

Von Glasersfeld, E.: 1991, 'Abstraction, re-presentation, and reflection: An interpretation of experience and Piaget's approach', in L. Steffe (ed.), *Epistemological Foundations of Mathematical Experience*, Springer-Verlag, New York, pp. 45–67.

Weiss, I., Boyd, S. and Hessling, P.: 1990, *A Look at Exemplary NSF Teacher Enhancement Projects*, Horizon Research, Inc, Chapel Hill, NC.

Wheeler, M. and Feghali, I.: 1983, 'Much ado about nothing: Preservice elementary school teachers' concept of zero', *Journal for Research in Mathematics Education* 14, 147–155.

Wilson P. and Padron, J.: 1994, 'Moving towards a culture-inclusive mathematics education', in M. Atwater, K. Radzik-Marsh and M. Strutchens (eds.), *Multicultural Education: Inclusion of all*, The University of Georgia, Athens, GA, pp. 39–63.

*University of Georgia,*
*College of Education – Mathematics Education,*
*Athens, GA 30602-7124,*
*U.S.A.*
*E-mail: TCooney@coe.uga.edu*

ANNA O. GRAEBER

# FORMS OF KNOWING MATHEMATICS: WHAT PRESERVICE TEACHERS SHOULD LEARN

ABSTRACT. What important ideas about forms of knowing mathematics should be included in mathematics methods courses for preservice teachers? Ideas are proposed that are related to categories in Shulman's (1986) framework of teacher knowledge. There is a brief discussion of the implications each idea holds for teaching mathematics, and some suggestions are given about experiences that may help preservice teachers appreciate these notions. One portion of Shulman's pedagogical content knowledge construct is knowing what makes a subject difficult and what preconceptions students are apt to bring. Three of the ideas offered for inclusion in a methods course are related to this aspect of pedagogical content knowledge: (1) Understanding students' understanding is important, (2) Students knowing in one way do not necessarily know in the other(s), and (3) intuitive understanding is both an asset and a liability. The last two ideas, are related to the other portion of pedagogical content knowledge, knowing how to make the subject comprehensible to learners. These ideas are (4) certain characteristics of instruction appear to promote retention, and (5) providing alternative representations and recognizing and analyzing alternative methods are important. Readers are asked to consider if the suggestions offered are appropriate and how they might best be taught.

## 1. INTRODUCTION

I considered the topic, forms of mathematical knowledge, from my perspective as a teacher educator engaged in teaching mathematics education courses and in developing teacher education programs. Thus this paper is an exploration of the question: What important ideas about forms of knowing mathematics should teachers meet in the mathematics methods portion of their preservice education?

Throughout discussions on knowing and learning, one notion is the need for clearly articulated overall ideas, principles, or 'big ideas.' Many authors describing learning consistent with constructivist views as well as authors of other persuasions have stressed this idea.

> ... the really useful training yields a comprehension of a few general principles within a thorough grounding in the way they apply to a variety of concrete details (Whitehead, 1929, p. 37).

> Structuring the curriculum around primary concepts is a critical dimension of constructivist pedagogy (Brooks and Brooks, 1993, p. 46).

*Educational Studies in Mathematics* **38:** 189–208, 1999.

Mathematics teachers desperately need to develop a network of insights and unifying concepts in the mathematics that they could possibly teach (Steffe, 1990, p. 184).

In reviewing some of our university students' work in mathematics, education, and in one case freshman chemistry, I was repeatedly struck by how difficult it was for college students to identify the 'big ideas' of a course. If we merely bemoan the fact that students do not seem to grasp these ideas, is it an instance of blaming the victim? Is it perhaps as Steffe's prescription and some of Tobias's work (1990) suggest, that faculty are unsure of what the main ideas might be or do a poor job of conveying them?

If we are to help future teachers reach some understanding about forms of knowing, teacher educators need to identify the central ideas we wish the preservice teachers to grasp. Below I propose such 'big ideas' or themes, outline some of the implications I see them as holding for teaching mathematics, and suggest some experiences that may help preservice teachers appreciate these notions.

## 2. What Big Ideas Should a Teacher Know About Knowing Mathematics?

What follows is by no means an attempt to list all of the ideas for a mathematics methods course. As Shulman (1986) and others have noted, there are many aspects of content knowledge, pedagogical content knowledge, and curricular knowledge. For Shulman, *content knowledge* is knowledge about the subject, for example mathematics, and its structure. *Pedagogical content knowledge* includes 'the ways of representing and formulating the subject that make it comprehensible to others' and 'an understanding of what makes the learning of topics easy or difficult; the concepts and preconceptions that students of different ages and backgrounds bring with them...' (Shulman, p. 9). Finally *curricular knowledge* encompasses what might be called the 'scope and sequence' of a subject and materials used in teaching. Methods courses typically address pedagogical content knowledge and curricular knowledge. In this paper I focus exclusively on some aspects of pedagogical content knowledge. In particular the discussion below is the result of thinking about important ideas about teaching and learning that often challenge the preservice teachers' intuitive ideas about teaching and learning. My attention is limited to issues related to forms of knowing mathematics as it concerns students' types of mathematical understanding and instructional strategies that promote student construction of meaning. The emphasis on construction of meaning comes from the perspective I take on learning, a cognitive/constructivist one.

Writing about preservice teachers also involves writing about teachers and students. This can become confusing. In this paper I use the term 'preservice teacher' to mean a person studying to teach mathematics as one of several subjects (generally future elementary teachers) or only mathematics (typically future secondary school teachers). Unless otherwise noted, the word 'student' is reserved for students from kindergarten to the end of secondary school. Although this paper is written about ideas for preservice teachers, it is my belief that all mathematics teachers should understand these ideas.

While not the major focus of the paper, some attention is also given to how one might actively involve teacher candidates in acquiring the knowledge, attitudes and skills suggested by each of these big ideas. It is acknowledged that the process of teacher education is itself complex and not well understood. The NCTM (1991) *Professional Standards for Teaching Mathematics* includes standards for the professional development of teachers of mathematics. In brief, these suggest that preservice teacher education ought to reflect the standards proposed for K-12 students. The Mathematical Sciences Education Board report (1996) notes that 'There is increased evidence that prospective teachers can learn about teaching mathematics from studying the 'practice of mathematics teaching' (p. 7). What is offered here are the author's suggestions that seem compatible with these ideas and that are based on what has proven feasible in her experience and context.

### 2.1. *Pedagogical content knowledge: instruction that promotes student construction of meaning*

In describing pedagogical content knowledge, Shulman included two large clusters of knowledge. One was knowing what makes understanding a particular concept difficult or easy and the conceptions or preconceptions students commonly bring to the frequently taught topics in the discipline. The second was knowing representations of regularly taught topics that would provide teachers with 'a veritable armamentarium of alternative forms of representation' (1986, p. 9).

In her article, 'What Constructivism Implies for Teaching,' Confrey (1990) notes that

> ... teachers must build models of student's understanding of mathematics ... through their interaction with students regarding their knowledge of subject matter, teachers construct a tentative path upon which students may move to construct a mathematical idea...' (p. 112).

This emphasis on understanding and building from students' current ideas is central to many who advocate a constructivist perspective (Brooks and

Brooks, 1993). The first three big ideas discussed below are included to highlight these concerns with following students' reasoning.

Shulman's framework, while attending to alternative representations and ways of making the subject comprehensible to others, does not reflect recent mathematics education emphases on eliciting from students various approaches to justifying a claim or understanding a concept. Similarly there is no discussion in his work of strategies that lead students to consider and reflect on their meaning in learning. The fourth and fifth big ideas discussed below describe strategies meant to make learning reasonable and memorable, and it seems logical that preservice teachers should have knowledge of them.

#### 2.1.1. *Understanding students' current understanding is important*

'Wrong' answers are usually not the result of carelessness, never having been taught, or lack of thought. 'Wrong' answers frequently have theoretical underpinnings that novice teachers do not anticipate. While merely telling the 'right idea' has not been found very successful in the face of student constructed knowledge, tapping into the student's existing ideas is frequently cited as the first step in helping students amend their notions (e.g., Driver, 1987; Swan, 1983). Understanding and supporting the student's reasoning are acknowledged by many as important to successful instruction. Thus it is important to understand children's understanding if you want them to amend or embellish what they know.

While preservice teachers typically study major learning theories (those of Piaget and Vygotsky, for example) in educational psychology classes, there are also some generally widespread patterns of thinking about a number of key school mathematics concepts that may be helpful for preservice teachers to know. For example researchers have identified 'stages' in the path to understanding place value (e.g., Steffe and Cobb, 1988) and theoretical hierarchies such as the van Hiele levels in geometry (Fuys, Geddes, and Tischler, 1988). There are also some hierarchies of the relative difficulty of ideas that seem to be useful in informing the sequencing of tasks. One of the clearest examples may be the research on types of addition and subtraction problems (e.g., join, separate, part-part-whole, compare) as delineated by Carpenter and Moser (1983). These schemes include notions of what is easy and difficult and fall into Shulman's category of pedagogical content knowledge. Of course, teachers must also be alert to the possibility that a student's actual knowledge may have ill-formed, fragile, or missing concepts that we normally presume to be at a 'lower level' than those concepts the student does hold.

While there has been some research (Putnam and Leinhart, 1986) that suggested that knowledge of students' thinking processes was not characteristic of 'expert teachers,' other research suggests the value of knowledge of students' thinking. Rine (1998) draws upon the work of The Wisconsin University's Cognitively Guided Instruction (CGI) Program and the University of California at Los Angeles' Integrating Mathematics Assessment Project (IMA). Both projects have presented evidence that students whose teachers learned about aspects of students' thinking about addition and subtraction word problems and fractions, respectively, increased their achievement in mathematics (Carpenter, Fennema, Peterson, Chiang and Loef, 1989; Gearhart, Saxe and Stipek, 1995). Such findings, Rine noted, suggest that increased achievement can be attained if teachers learn about students' thinking in a variety of topics. He also notes the difficulty of this implication; learning all this may simply be too much to ask of teachers. Rine suggests that teachers might best learn to assess students' thinking during instruction.

Clearly preservice teachers cannot learn all that is known about students' thinking on a wide range of mathematical topics. However, it does seem reasonable to ask that preservice teachers know what is understood about students' thinking in some areas (and know that such knowledge exists in other areas) and understand how assessing students' thinking can give direction to instruction.

*Implications for teaching mathematics.* General patterns of development are useful in designing experiences and recognizing where a student may be in relation to some scheme. Nevertheless, classroom instruction must involve ongoing assessment of students' understanding of concepts. Geometry teachers often assume that students know what a square or a right triangle is. Yet there is evidence that given such a figure in a nonstandard position many high school students ready to enter a geometry course have difficulty identifying them (e.g., Hershkowitz, Bruckheimer and Vinner, 1987). This awareness of individual concepts is a tall order for the classroom teacher and helping teachers find means for doing this and using the information to shape instruction have been areas of interest among mathematics education researchers (e.g., those involved in the Cognitively Guided Instruction Program, see for example Fennema et al., 1996).

Helping preservice teachers realize the usefulness of such ongoing assessment and hierarchies for making instructional decisions would seem to be important. If preservice teachers enter the classroom without valuing student understanding, they are not apt to assess understanding or use knowledge of students' current understanding to make instructional decisions.

*Build preservice teachers' appreciation.* I have found that only by combining reading about common misconceptions and limited conceptions with interviewing or tutoring experiences are preservice teachers convinced of the need to understand students' understanding. The readings alert them to potential misconceptions. The tutoring experience gives the preservice teachers a chance of meeting a student with such a misconception. And, it is helpful to be able to witness (or videotape) the preservice teacher's tutoring and help them reflect on some of what they see or hear. I find that often in their attempt to 'deliver' a lesson, they disregard evidence of a student's lack of understanding or misunderstanding. Whether this is done consciously or subconsciously to 'get on with' a smooth lesson, or whether the preservice teachers simply do not hear evidence that signals misunderstanding, I do not know. I do know from observing numerous tutoring sessions that they often allow signs of misunderstanding to simply go by. Nevertheless, my sense is that it is more likely that preservice teachers will detect a student's misconception in a tutoring situation than in a large class situation where the novice is also highly absorbed in the management of the class.

In one of the tutoring sessions a preservice teacher was working with a seventh grader who was having problems reading decimal numerals. The student read 0.6 as six tenths, 0.57 as fifty-seven tenths and 0.039 as thirty-nine hundredths. The preservice teacher first took out a sheet of paper, constructed a place-value chart and asked the student to name the places to the right of the decimal point. The student did this correctly. The preservice teacher then asked the student how he named decimals. The student replied 'Well, you name them by where they start, don't you?' The methods class discussion of this episode pointed out the inefficiency of not initially asking about the student's understanding and the advantage of doing so.

Also useful are videotapes of interviews or instruction that demonstrate the power and limits of student's knowledge. A 1993 M. Burns video-tape shows a student who can count out 16 chips, can associate six counters with the digit 6, but then indicates that the 1 in 16 stands for one chip. The student persists in this even when pressed for other meanings and asked to explain the meaning of the remaining 9 (16 – 6 – 1) chips. The student's performance suggests that he possesses only a 'unitary' concept of number. Steffe and Cobb's (1988), Fuson's (1990) and other's work indicates that this is a common stage of understanding in the development of place value concepts. The research also indicates that the student is unlikely to understand a ten ones for one ten trade. Teaching such a student a multi-digit addition algorithm, while not necessarily inappropriate, requires careful thought about the student's potential for understanding. Attempts at simply explaining the written algorithm are apt to require understanding of numerals this student does not yet have. If manipulatives are to be used to support understanding of a written algorithm, even the choice of manipulative might be a consideration here. Should the student work with structured materials such as base ten blocks or unstructured materials such as linkable cubes? Are the prestructured blocks appropriate or would the linkable materials that allow the student to physically construct groups of ten from ones be more appropriate? While the student's mental conceptions are the ultimate concern, helping the student build the notion of ten ones for one ten might involve connecting a physical activity with a written scheme. These are the types of pedagogical problems which with preservice teachers will face. If preservice teachers understand that instructional decisions can be guided by what is known about children's understanding, they may be more motivated to pursue understanding of the children's understanding.

#### 2.1.2. *Students knowing in one way do not necessarily know in the other(s)*

Not only must preservice teachers be aware of students' thinking, but they must be skillful in distinguishing what the students understand as opposed to what the students can do.

'The relationship between computational skill and mathematical understanding is one of the oldest concerns in the psychology of mathematics' (Resnick and Ford, 1981, p. 246). In general, the terms skill and understanding distinguish between what is merely memorized, can be recited or performed, and what can be applied in various contexts.

In his *Aims of Education* (1929), Whitehead distinguishes between inert ideas, 'ideas that are merely received into the mind without being utilized, or tested, or thrown into fresh combinations,' (p. 13) and those a learner can make his own and apply in life.

Skemp summarizes the distinctions between instrumental and relational learning as knowing how as opposed to knowing both how and why. His discussion of the advantages and disadvantages of each type of learning and results of teachers and students with differing objectives is a very helpful resource (1978).[1]

Hiebert and LeFevre (1986) proposed their definition of conceptual and procedural knowledge recognizing that the distinctions they drew were not always consistent with past uses of the terms. They argued that 'it is the *relationships* between conceptual and procedural knowledge that hold the key' (p. 23).

Fischbein (1993) describes algorithmic, formal and intuitive dimensions of knowledge. The algorithmic is basically procedural; and the formal includes definitions of concepts, operations, structures and the axioms relevant to some domain. Fischbein's notion of the intuitive includes those dominant ideas, beliefs and models that seem to be obvious and in no need of proof.

Different theories of learning or epistemologies emphasize, or accept as evidence of understanding, different forms in different mixes. At different times in the history of education in a country, different forms of understanding may be more popular than others. For example, Howson, Keitel and Kilpatrick (1981) point out that although teaching mathematics for understanding was still a concern of many mathematics educators in the United States in the 1970's, the back-to-basics era was characterized by widespread emphasis on knowing basic facts and employing principles of behaviorist theory in engineering curriculum. The behaviorists generally had no concern for evoking the reasoning that led to a result. In recent years, there have been lively debates in both the United States and Great Britain about the relevant importance of computation and understanding (e.g., Berger and Keynes, 1995; London Mathematical Society, 1995).

The forms of knowing deemed important at a particular time shape goals as reflected in curriculum and in recommended strategies for instruction. They shape what and how we assess learning. As Skemp (1978) articulated so well, students' views and teachers' views of what forms of knowing are important may differ and cause conflict in goals. It would seem helpful for teachers to know which practices are consistent with and support the outcomes of varied goals.

Prospective teachers must understand that students' who posses one form of knowledge do not necessarily possess other forms of knowledge. For example, students may hold procedural knowledge of how to multiply two common fractions, but have poor conceptual knowledge of either fractions or multiplication. Or students may understand the notion of multiplication of whole numbers as repeated addition, but not possess a means of finding such a product without adding.

People can also hold conflicting ideas in the different forms. Graeber and Tirosh (1988), for example, noted that 87% of a group of preservice teachers correctly executed a division algorithm in which the divisor was a decimal less than one. However, only 45% of the same 136 preservice teachers identified as false the statement, 'In a division problem, the quotient must be less than the dividend,' (p. 270). Thus a non-trivial number of the preservice teachers knew how to compute a dividend, but their reactions to explicit statements were not consistent with their own computation.

*Implications for teaching mathematics.* It seems to me important for teachers to understand that executing an algorithm, or getting the right answer does not imply conceptual understanding. This understanding is at the heart of the argument that new goals must be accompanied by new forms of assessment (Webb and Romberg, 1992; Bell, Burkhardt and Swan, 1992). If assessments only evoke procedural understanding, students' concepts are not explored. Similarly, conceptual understanding does not always assure computational facility. If conceptual and procedural understanding are goals, teachers must select tasks that evoke understanding as well as execution.

And, since students will often not recognize their own inconsistent ideas as inconsistent, teachers must be alert to the possibility and be able to create tasks or situations that foster recognition and amendment of such conflicting ideas. Different authors have suggested different approaches to helping students gain consistent ideas, Bell, Brekke and Swan's conflict teaching method (1987), Clement and others' bridging strategy (Brown and Clement, 1989), and Fischbein's (1987) ideas about developing and utilizing secondary intuitions.

If preservice teachers enter the classroom without making the distinction between conceptual and procedural knowledge or without recognizing that one type of knowledge does not necessarily imply the other, they are apt to take existence of one type as evidence of existence of the other.

*Build preservice teachers' appreciation.* Preservice teachers can interview or tutor students, requesting both computation and pressing them for definitions, characteristics, and estimates. Often this will lead the aspiring teachers to see that the conceptual and procedural forms can exist independently. I find that such actual experiences are more convincing than having students simply read about or even view videotapes of illustrative instances. Numerous times I have seen preservice teachers surprised that students can verbalize and express confidence in numerous equivalent fractions when viewing a model such as Fraction Circles™ but reject written expressions such as $\frac{1}{3} = \frac{4}{12}$.

University instructors who cannot provide tutoring experiences for preservice teachers can use videos of students who clearly demonstrate a student holding only one form of understanding. For example in the 1993 videotape mentioned above, Marilyn Burns interviews a student who can make and identify two sets of tens and four left over counters, but who cannot tell how many counters there are until she counts them one-by-one. Preservice teachers can be asked to characterize the forms of knowing evoked and identify what concepts have not been attained or connected to other concepts.

Often there are computations the preservice teachers themselves can do without a conceptual understanding. They can frequently divide two common fractions but cannot give a rationale for the size of the answer or construct a real world interpretation of such a problem. Once familiar with the notions of conceptual and procedural knowledge, preservice teachers often can recognize this state (knowing how, but not why) in their own learning of college level mathematics. As Hiebert and Wearne (1986) have pointed out, not having a conception of an operation such as multiplying decimals and some number sense deprives students of the ability to estimate answers. In the case of division of fractions, conceptual understanding and number sense will help students decide which number should be inverted (i.e., is $\frac{2}{3} \div \frac{3}{4}$ the same as $\frac{3}{2} \times \frac{3}{4}$ or $\frac{2}{3} \times \frac{4}{3}$?) when dividing fractions. The fact that $\frac{3}{4}$ is larger than $\frac{2}{3}$ together with a measurement concept of division should lead to a quotient less than one. Thus $\frac{2}{3} \times \frac{4}{3}$ is an appropriate calculation for $\frac{2}{3} \div \frac{3}{4}$. This is a form of mathematical power that conceptual understanding can bring and that some preservice teachers can appreciate.

### 2.1.3. *Intuitive knowledge is both an asset and a liability*

Tasks that are designed to evoke conceptual knowledge, often evoke intuitive knowledge. (Intuitive knowledge is used here in the sense Fischbein does – immediate, self-evident knowledge (Fischbein, 1987, p.6)). The literature on misconceptions contains a plethora of examples in which intuitive mathematical knowledge differs from the knowledge accepted by the mathematical community. Multiplication, considered as repeated addition, is helpful in solving problems such as 'If 1 pound of coffee costs \$6.50, what is the cost of 2 pounds of coffee?' (I can add \$6.50 twice.) However, this same intuitive understanding of multiplication is not helpful if we ask 'If 1 pound of coffee costs \$6.50, what is the cost of 0.2 pounds of coffee?' (How do I add \$6.50 two-tenths of a time? If multiplication is repeated addition, how can multiplication give me an answer smaller than \$6.50?) Intuition plays a key role in problem solving or in construction of a proof, where students must take first steps based on what they *feel* are similarities to other problems, a visual model of the problem, or the start of a string of logic that will lead to the desired conclusion. On the other hand the certitude that accompanies many intuitive notions often leads to blind alleys or years of acceptance of ideas as certain, e.g. Euclid's fifth postulate (Fischbein, 1987, p. 63).

When teaching for understanding, teachers are apt to elicit students' preconceptions. Shulman noted 'if those preconceptions are misconceptions, which they so frequently are, teachers need knowledge of the strategies most likely to be fruitful in reorganizing the understanding of learners, because those learners are unlikely to appear before them as blank slates' (1986, pp. 9–10). And, as Fischbein (1987) argued, there is a need both to preserve students' confidence in their intuition and at the same time help them develop a healthy skepticism of intuitive ideas. Intuition is useful in all real problem solving, yet as the literature on misconceptions illustrates, intuition can often lead to erroneous conclusions.

*Implications for teaching mathematics.* Fischbein (1987) noted the importance of sharing with students the logic of their intuitive yet faulty beliefs such as 'multiplication always makes bigger.' He argued that the student must understand that others share this (faulty) belief and that there is a logic to such a belief. At the same time the student must see a value in the risk taken in following one's intuition (see Fischbein, 1987, pp. 36–42). Driver (1987) has also suggested that the classroom conditions needed to facilitate conceptual change include the providing of a non-threatening environment where students feel comfortable in expressing and sharing their views.

This dual nature of intuition suggests that teachers should help students develop the 'habit of mind' of challenging the 'obvious' and seeking alternative ways of confirming what seems 'certain.' If preservice teachers enter the classroom without these understandings, they may either let naive conceptions go unchallenged or they may so undermine students' confidence in their own intuition that the students' problem solving is hampered by their belief that all intellectual 'leaps' will lead to erroneous ideas.

*Build preservice teachers' appreciation.* Reading about and discussing a number of prominent misconceptions can be one way to help preservice teachers realize the models, curriculum experiences, and use of language that promote misconceptions such as 'division always makes smaller, to add fractions add the numerators and add the denominators, a square is not a rectangle, the more digits to the right of a decimal point the smaller the number, you need to have a mathematical mind to do school mathematics' etc. (Such reading and discussion are also a gentle way of helping some preservice teachers discover their own misconceptions.) Preservice teachers often think that such ideas can simply be 'told' away. Actual teaching/tutoring experience is often needed to amend this notion (misconception).

Sometimes I find it useful to provide problem solving experiences that are apt to result in some preservice teachers' intuition leading them to erroneous or cumbersome solutions. I have given secondary education majors the following stereotypic problem:

Chris sails upstream at 3 km/hour and downstream at 9 km/hour. If she starts upstream at 10 AM at what time should she turn around in order to be back at her starting point at 2 PM the same day?

For many the problem triggered an algebraic solution using a variable and the t = d/r formula. Others simply noted that the time must be in a ratio of 3:1 with three of the time units spent going upstream, hence in the four hours available, Chris can spend three going upstream.

The problem itself is admittedly of little 'real life' interest. But it was used to raise the notion that alternative strategies are sometimes more efficient than routine application of a well established method. The discussion that ensued encompassed a number of concerns. Could the students who used the ratio method justify their reasoning? Could those who chose the algebraic method note what made them take that path? Were both methods valid? In using such exercises, I find the instructor must remember Fischbein's admonitions about caring for the preservice teachers' intuition. Some preservice teachers find alternative methods a threat to what has worked for them, and it may raise self-doubts about their ability to deal with students alternative solution methods.

2.1.4. *Certain instructional characteristics appear to promote retention*
Those who argue for teaching for understanding, generally claim advantages for meaningful learning. These include, among others, enhancing the likelihood of remembering (Hiebert and Carpenter, 1992, p. 74). If students are to remember the mathematics they are taught, teachers must understand strategies that will promote remembering. There is evidence to suggest that what has been learned in a meaningful context, through reasoning from relatively primitive concepts, that has been explicated to others, and that was considered by reflecting on one's own change in knowledge is apt to be remembered.

Familiar contexts are found helpful in assisting problem solving as well as in motivating the students. This is part of the rationale underlying current curriculum attempts to provide problem settings prior to giving algorithmic procedures. These ideas are at the heart of discussions about situated cognition (Brown, Collins and Duguid, 1989).

Reasoning from primitive concepts is an empowering strategy. An example in the literature is the work by Wearne and Hiebert on learning the 'rules' for operations with decimals from understanding the meaning of decimals (1988).

Writing or speaking about or in other ways reflecting on knowledge and change in knowledge is considered by many an important strategy in learning. Theory and evidence about writing in the mathematics classroom can be found in Connolly and Vilardi (1989) and in Ellerton and Clements (1991). An eloquent brief for student discussion in the classroom is found in Yackel, Cobb, Wood, Wheatley and Merkel (1990).

Bell, Brekke, Swan (1987) and Driver (1987) are among those who have argued the benefits of having students look at their own changes in thinking by looking back at what they used to think and comparing it with what they think after an instructional sequence.

*Implications for teaching mathematics.* The research cited above and other similar findings provide a rationale for requiring students to explain their thinking and justify their procedures and answers to meaningful problems. It suggests a classroom where teachers do less talking and more listening to students' responses and to students' discussions with one another. Teachers who have attempted to implement such classrooms find that they need to teach children how to participate in such classrooms, learn different strategies to evoke student's reasoning, and maintain a focus on important mathematics (see for example, Silver and Smith, 1996). How one helps teachers develop these implementation skills is beyond the scope of this paper. Here, the concern is with building preservice teachers awareness of the instructional strategies.

If preservice teachers are not familiar with the advantages of and examples of such instructional principles, they are apt to repeat the type of 'teaching [only] by telling' that has characterized so much mathematics instruction in the past. If one takes teaching for understanding seriously, the literature on conceptual change suggests that 'teaching [only] by telling' will not result in the kind of understanding that most wish for students (see for example, Mestre, 1987).

*Build preservice teachers' appreciation.* Preservice teachers need time both to experience these activities as learners and to design lessons that incorporate these notions for their students. For example, helping preservice teachers see how the definition of common fraction can be used in ordering fractions and determining fractions near benchmarks of 0, $\frac{1}{2}$, and 1 can help them appreciate the power of primitive concepts. (Interpreting

$\frac{1}{10}$ as one of ten same size parts of the whole, should lead to the realization that $\frac{1}{10}$ is a small portion of the whole. Interpreting $\frac{6}{10}$ as six of ten same size parts of a whole, should result in the realization that $\frac{6}{10}$ is close to, but somewhat more than, $\frac{1}{2}$. Similarly interpreting $\frac{9}{10}$ as nine of ten same size parts of a whole, should signal that the fraction $\frac{9}{10}$ is close to but less than one.) Engaging the preservice students in writing to explain their reasoning in solving a problem can help them reflect on their own thinking.

Requiring preservice teachers to talk with a student so they know the contexts (hobbies, jobs, interests) that would be meaningful to the students they tutor can be a first step in encouraging the preservice students to use such contexts. I ask the preservice teachers to plan lessons that build from conceptual understanding to a procedure or generalization. The lesson plan might require the student to write a summary of his or her learning, thereby giving the preservice teacher insight into what they can learn from student's writing. Asking preservice teachers to interview students and report on a student's understanding encourages listening to student ideas.

Occasionally there are fortuitous events in tutoring which illustrate some of these ideas. Toward the end of the last semester my preservice teachers were encouraged to challenge their middle school tutees with several items from the recent Voluntary National Test of Mathematics (VNTM)[2]. One VNTM item was as follows. *Luis exercises by running 5 miles each day. The course or track he runs is $\frac{1}{4}$ mile long. How many times does he run the course each day?* (MPR Associates and Chief State School Officers, 1997, p. 104.). A preservice teacher elected to give this item to a special education student that had had great difficulty with fractions, equivalent fractions, etc. The student solved the problem with great success, noting that four laps was a mile, etc. The preservice teacher was amazed at the middle school student's success and asked how he had come to his answer with such speed and confidence. The student noted that he ran track, so it was very clear to him how to do the problem. The class was able to view this as an instance of meaningful context making what one might consider to be a formidable problem not so difficult.

### 2.1.5. *Alternative representations and the recognition and analysis of alternative methods are important*

Encapsulated in this brief title are a number of related ideas: there are different logical or experiential paths that lead to the same ideas; similar experiences may lead to different yet valid ideas; different models help different students construct ideas, different students make different connections and different numbers of connections of ideas. These are neither new nor startling ideas. They are included here because they relate to the variety of approaches students may use in formulating knowledge of mathematics. Thus there is an obvious connection to preservice teachers' pedagogical content knowledge related to alternative representations and success in helping students achieve conceptual understanding.

*Implications for teaching mathematics.* If all students are to learn, multiple approaches need to be encouraged. While there have been many studies and debates about the number of alternative representations, how they might be ordered, and what conditions are needed to promote the effectiveness of different representations; the conviction remains that representations can be useful in helping individuals construct and add to relationships (Hiebert and Carpenter, 1992, pp. 69–72). However, the problem-first, strategy-development-and-practice later approach of many newer curricula demands even more than simply the exposition of different models. They require that teachers be able to help students recognize the validity, generalizability and efficacy of the students' own solution methods. Methods that may not be those the teacher expected. Such teaching places demands not only on content knowledge and knowledge of representations but also on the teachers' attitudes toward teaching and learning.

If preservice teachers fail to provide alternative paths to understanding, they are apt to leave some students without understanding. If they fail to recognize and analyze alternative solutions to problems, students' reasoning may be undervalued or, more seriously, be declared incorrect if valid or correct when invalid.

*Build preservice teachers' appreciation.* Preservice teachers, especially those in secondary mathematics education, are often quick to note that the old way – homework review, lecture, practice– worked for them, why do they have to do anything else such as approach ideas through problem solving or applications or use various models? They need to develop a new attitude toward teaching and learning. Asking them to consider just what percent of the population the 'old way' worked for sometimes gives them pause to consider new ideas.

Clearly knowing different models and various approaches to topics places demands on the preservice teachers' mathematical knowledge as well as on their attitude. Brophy has noted the following about teacher's subject matter knowledge

> Where [teachers'] knowledge is more explicit, better connected, and more integrated, they will tend to teach the subject more dynamically, represent it in more varied ways, and encourage and respond fully to student comments and questions. (Brophy, 1991, p. 352).

Again I think preservice teachers need to experience the notion of multiple approaches in the methods classroom and be required to plan for multiple approaches with their students.

In the methods class the preservice teachers can be asked to solve problems with a model, not with computation. The discussion can then include issues such as: What did they find helpful in different models of signed numbers? What was not helpful? In modeling division of fractions, what did the Fraction Factory™ (unit represented by a rectangle) manipulative allow that the Fraction Circle™ (unit represented by a circle) did not? What were the disadvantages of the circular model, or the rectangular model?

In one of my classes I engaged preservice teachers in generalizing from examples of subtraction of signed integers using a two-color chip model. One student posited the following: 'If I am subtracting two integers with unlike signs, the sign of the difference is the same as the sign of the minuend (or sum).' A discussion of the validity and value of the statement ensued. How would they 'prove' to themselves and to the students that the statement was in fact valid? How did the rule compare to the oft repeated 'To subtract, change the sign of the bottom number and follow the rules for addition' or the more formal 'Subtracting a number is the same as adding its opposite?'

Requiring preservice teachers to write a lesson plan (or plans) that utilizes different models for the same concept and encourages students to link the models to the mathematics and to one another can be a way of challenging preservice teachers to envision alternative approaches.

## 3. Are These the Right Big Ideas? How Do Preservice Teachers Lean Them?

Just as students do not master the complexity of ideas such as place value in one lesson or one year, preservice teachers are not apt to master these five ideas in one course. Nevertheless to send them into the classroom without serious attempts to help them own these ideas, seems to me to be doing the profession a disservice.

There are also many other important concepts for a methods course. Many of which are outlined in writings such as Shulman (1986) and NCTM (1991). But there are always far more things that would be worthwhile for a methods course than there is time. Within the admittedly narrow domain considered in this article, forms of knowing mathematics, I would like to see continued discussion of and research on the following issues:

1) Are the 'big ideas' about forms of knowing listed the important ones for preservice teachers? Are they our best understanding of the most important?
2) What are effective ways of helping preservice teachers to know how, know when, and know-to utilize this knowledge? In other words, how do prospective teachers learn ideas about pedagogical knowledge and how do they learn to apply them?

## NOTES

1. Skemp's definition of relational suggests that students with relational understanding also have instrumental understanding.
2. In 1997, U.S. President Clinton called for a voluntary national test in 8th grade mathematics. While such testing has not been implemented, items and item specifications were published by MPR Associates and The Chief State School Officers (1997).

## REFERENCES

Bell, A., Brekke, G. and Swan, M.: 1987, 'Misconceptions, conflict and discussion in the teaching of graphical interpretation', in J. Novak (ed.), *Proceedings of the Second International Seminar on Misconceptions and Educational Strategies in Science and Mathematics*, Vol 1, Cornell University, Ithaca, New York, pp. 46–48.

Bell, A., Burkhardt, H. and Swan, M.: 1992, 'Balanced assessment of mathematical performance', in R. Lesh and S. Lamon (eds.), *Assessment of Authentic Performance in School Mathematics*, American Association for the Advancement of Science, Washington, DC, pp. 119–144.

Berger, T. and Keynes, H.: 1995, 'Everybody counts/everybody else', *CBMS Issues in Mathematics Education* 5, 89–110.

Brooks, J. and Brooks, M.: 1993, *In Search of Understanding: The Case for Constructivist Classrooms*, Association of Supervision and Curriculum Development, Alexandria, Virginia

Brophy, J. E.: 1991, 'Conclusion to advances in research on teaching', in J. Brophy (ed.) *Advances in Research on Teaching: Teachers' Subject-matter Knowledge and Classroom Instruction: VOL. II Teachers' Knowledge of Subject Matter as it Relates to Teaching Practice*, JAI Press, Greenwich, Connecticut, pp. 347–362.

Brown, D. E. and Clement, J.: 1989, 'Overcoming misconceptions by analogical reasoning: Abstract transfer versus explanatory model construction', *Instructional Science* 18, 237–261.

Brown, S., Collins, A. and Duguid, P.: 1989, 'Situated cognition and the culture of learning,' *Educational Researcher* 18(1), 32–42.

Burns, M.: 1993, *Mathematics: Assessing and Understanding. Individual Assessments, Part I.* [Videotape]. Available from Cuisenaire/Dale Seymour Publications, P.O. Box 5026, White Plains, New York 10602.

Carpenter, T., Fennema, E., Peterson, P., Chiang, C., and Loef, M.: 1989, 'Using knowledge of children's mathematics thinking in classroom teaching: An experimental study', *American Educational Research Journal* 26, 499–531.

Carpenter, T. and Moser, J.: 1983, 'The acquisition of addition and subtraction concepts', in R. Lesh and M. Landau (eds.), *The Acquisition of Mathematics Concepts and Processes,* Academic Press, New York, pp. 7–44.

Confrey, J.: 1990, 'What constructivism implies for teaching', in R. B. Davis, C. A. Maher and N. Noddings (eds.), *Constructivist Views on the Teaching and Learning of Mathematics*, National Council of Teachers of Mathematics, Reston, Virginia, pp. 107–122.

Connolly, P. and Vilardi, T.: 1989, *Writing to Learn Mathematics and Science*, Teachers College Press, New York.

Driver, R.: 1987, 'Promoting conceptual change in classroom settings: The experiences of the children's learning science project', in J. Novak (ed.), *Proceedings of the Second International Seminar on Misconceptions and Educational Strategies in Science and Mathematics*, Vol 2, Cornell University, Ithaca, New York, pp. 97–107.

Ellerton, N. and Clements, M. A.: 1991, *Mathematics in Language: A Review of Language Factors in Mathematics Learning*, Deakin University Press, Geelong, Australia.

Fennema, E., Carpenter, T., Franke, M., Levi, L., Jacobs, E. and Empson, S.: 1996, 'A longitudinal study of learning to use children's thinking in mathematics instruction', *Journal for Research in Mathematics Education* 27(4), 403–434.

Fischbein, E.: 1993, 'The interaction between the formal, the algorithmic and the intuitive components in a mathematical activity', in R. Biehler, R. Scholz, R. Straser, and B. Winkelmann (eds.), *Didactics of Mathematics as a Scientific Discipline*, Kluwer, Dordrecht, The Netherlands, pp. 231–245.

Fischbein, E.: 1987, *Intuition in Science and Mathematics*, D. Reidel, Dordrecht, The Netherlands.

Fuson, K.: 1990, 'Issues in place-value and multidigit addition and subtraction learning and teaching,' *Journal for Research in Mathematics Education* 21(4), 273–279.

Fuys, D., Geddes, D. and Tischler, R.: 1988, *The van Hiele Model of Thinking in Geometry Among Adolescents* (Journal for Research in Mathematics Education Monograph #3), National Council of Teachers of Mathematics, Reston, Virginia.

Gearhart, M., Saxe, G. B. and Stipek, D.: Fall 1995, 'Helping teachers know more about their students: Findings from the Integrating Mathematics Assessment (IMA) project', *Connections* (1), 4–6, 10.

Graeber, A. and Tirosh, D.: 1988, 'Multiplication and division involving decimals: Preservice elementary teachers' performance and beliefs', *The Journal of Mathematical Behavior* 7(3), 263–280.

Hershkowitz, R., Bruckheimer, M. and Vinner, S.: 1987, 'Activities with teachers based on cognitive research', in M. Lindquist (ed.), *Learning and Teaching Geometry, K-12,* 1987 Yearbook, National Council of Teachers of Mathematics, Reston, Virginia, pp. 222–235.

Hiebert, J. and Carpetner, T.: 1992, 'Learning and teaching with understanding,' in D. Grouws (ed.), *Handbook of Research on Mathematics Teaching and Learning*, MacMillan, New York, pp. 65–97.

Hiebert, J. and Lefevre, P.: 1986, 'Conceptual and procedural knowledge in mathematics: An introductory analysis', in J. Hiebert (ed.), *Conceptual and Procedural Knowledge: The Case of Mathematics*, Lawrence Erlbaum, Hillsdale, New Jersey, pp. 1–27.

Hiebert, J. and Wearne, D.: 1986, 'Procedures over concepts: The acquisition of decimal number knowledge', in J. Hiebert (ed.), *Conceptual and Procedural Knowledge: The Case of Mathematics*, Lawrence Erlbaum, Hillsdale, New Jersey, pp. 199–224.

Howson, G., Keitel, C. and Kilpatrick, J.:1981, *Curriculum Developments in Mathematics*, Cambridge University Press, Cambridge, England.

London Mathematical Society, Institute of Mathematics and Its Applications and Royal Statistical Society: October, 1995, *Tackling the Mathematics Problem.*

Mathematics Science Education Board: 1996, *The Preparation of Teachers of Mathematics: Considerations and Challenges, A Letter Report*, National Research Council, Washington, DC.

Mestre, J.: 1987, 'Why should mathematics and science teachers be interested in cognitive research findings?', *Academic Connections* 3–5, 8–11.

MPR Associates and The Chief State School Officers: 1997, *Item and test specification for the voluntary national test in 8th-grade mathematics*, MPR Associates, Washington, DC.

National Council of Teachers of Mathematics: 1991, *Professional Standards for Teaching of Mathematics*, The Council, Reston, Virginia.

Putnam, R. T. and Leinhardt, G.: 1986. 'Curriculum scripts and the adjustment of content in mathematics lessons', paper presented at the annual meeting of the American Educational Research Association, San Francisco, California.

Resnick, L. and Ford, W.: 1981, *The Psychology of Mathematics Instruction*, Lawrence Erlbaum, Hillsdale, New Jersey.

Rine, S.: 1998, 'The role of research and teachers' knowledge base in professional development,' *Educational Researcher* 27(5), 27–31.

Shulman, L. S.: 1986, 'Those who understand: Knowledge growth in teaching', *Educational Researcher* 15(2), 4–14.

Silver, E. A. and Smith, M. S.: 1996, 'Building discourse communities in mathematics classrooms: A worthwhile but challenging journey', in P. Elliot (ed.), *Communication in Mathematics, K-12 and Beyond*, 1996 Yearbook, National Council of Teachers of Mathematics, Reston, Virginia, pp. 20–28.

Skemp, R.: 1978, 'Relational understanding and instrumental understanding', *Arithmetic Teacher* 26(3), 9–15.

Steffe, L.: 1990, 'On the knowledge of mathematics teachers', in R. B. Davis, C. Maher, and N. Noddings (eds.), *Constructivist Views on the Teaching and Learning of Mathematics*, JRME Monograph #4, National Council of Teachers of Mathematics, Reston, Virginia, pp. 167–184.

Steffe, L. and Cobb, P.: 1988, *Construction of Arithmetical Meanings and Strategies*, Springer-Verlag, New York.

Swan, M.: 1983, *Teaching Decimal Place Value: A Comparative Study of 'Conflict' and 'Positive Only' Approaches*, Shell Centre for Mathematics Education, University of Nottingham, Nottingham, England.

Tobias, S.: 1990, 'They're not dumb. They're different: A new "Tier of Talent" for science', *Change* 22(4), 10–30.

Wearne, D. and Hiebert, J.: 1988, 'A cognitive approach to meaningful mathematics instruction: Testing a local theory using decimal numbers', *Journal for Research in Mathematics Education* 19(5), 371–384.

Webb, N. and Romberg, T.: 1992, 'Evaluation a coat of many colors', in T. Romberg, (ed.), *Mathematics Assessment and Evaluation*, State University of New York Press, Albany, New York, pp. 10–36.

Whitehead, A.: 1929, *The Aims of Education*, MacMillan, New York.
Yackel, E., Cobb, P., Wood, T., Wheatley, G. and Merkel, G.: 1990, 'The importance of social interaction in children's construction of mathematical knowledge,' in T. Cooney and C. Hirsch (eds.), *Teaching and Learning Mathematics in the 1990s*, 1990 Yearbook, National Council of Teachers of Mathematics, Reston, Virginia, pp. 12–21.

*University of Maryland,*
*Department of Curriculum and Instruction,*
*2311 Benjamin Building,*
*College Park, MD 20742,*
*U.S.A.*
*E-mail: ag20@umail.umd.edu*

PESSIA TSAMIR

# THE TRANSITION FROM COMPARISON OF FINITE TO THE COMPARISON OF INFINITE SETS: TEACHING PROSPECTIVE TEACHERS

ABSTRACT. Research in mathematics education indicates that in the transition from given systems to wider ones prospective teachers tend to attribute all the properties that hold for the former also to the latter. In particular, it has been found that, in the context of Cantorian Set Theory, prospective teachers have been found to erroneously attribute properties of finite sets to infinite ones – using different methods to compare the number of elements in infinite sets. These methods which are acceptable for finite sets, lead to contradictions with infinite ones. This paper describes a course in Cantorian Set Theory that relates to prospective secondary mathematics teachers' tendencies to overgeneralize from finite to infinite sets. The findings clearly indicate that when comparing the number of elements in infinite sets the prospective teachers who took the course were more successful and were also more consistent in their use of a single method than those who studied the traditional, formally-based Cantorian Set Theory course.

## 1. Introduction

In the course of accumulating mathematical knowledge, the student goes through successive processes of generalization, while also experiencing the extension of various mathematical systems. For instance, the concept of number, a central concept in mathematics, is introduced very early on in primary school via the system of natural numbers. Then, gradual transitions occur beginning with integers, through rational numbers, irrational numbers and real numbers, and concluding with the system of complex numbers presented in the upper grades of high school. The move from one number system to a wider one adds some numerical characteristics while some other attributes are lost. For example, the transition from natural numbers to integers enables one to solve a problem like 5–7 (closure under subtraction). Yet, at the same time it becomes impossible to generalize that addition 'always makes bigger' and the system no longer has a smallest number.

Awareness of the changes caused by the enlargement of mathematical systems and the ability to identify the variant and invariant elements under a specific transition, are important factors in the growth of mathematical

*Educational Studies in Mathematics* **38:** 209–234, 1999.

knowledge. However, research findings clearly indicate that students and teachers, pre-service as well as in-service, tend to attribute *all* properties of a specific domain of numbers, to a more general one (e.g., regarding rational numbers: Greer, 1994; Hart, 1981; Klein and Tirosh, 1997; regarding decimals: Moloney and Stacey, 1996; Putt, 1995; regarding negative numbers: Hefendehl, 1991; Streefland, 1996; regarding irrational numbers: Fischbein, Jechiam and Cohen, 1995; and regarding complex numbers: Almog, 1988).

The transition from the comparison of finite sets to the comparison of infinite sets (i.e., the comparison of the number of elements in these sets) is another example of an extension of a mathematical system. Research in the field of mathematics education indicates that this transition is problematic for many students (e.g., Duval, 1983; Falk, Gassner, Ben Zoor and Ben Simon, 1986; Fischbein, Tirosh and Melamed, 1981; Martin and Wheeler, 1987; Tsamir and Tirosh, 1992, 1994). When asked to compare the number of elements in two infinite sets, students used different methods in their comparisons, leading to unavoidable contradictions. For instance, when asked to compare the number of elements in the set of natural numbers and the set of positive, even numbers, students used both inclusion and 1:1 correspondence as their bases for comparison. The use of *inclusion* (i.e., justifying the claim that two sets consist of different numbers of elements by stating that a set included in another set has less elements), led them to conclude that the infinite sets have an unequal number of elements. When using *1:1 correspondence* (i.e., justifying the claim that two sets have the same number of elements by pairing each element of one set with a unique element of the other), they concluded that the number of elements was equal. Students often accepted these two contradictory solutions to the same problem as valid (e.g., Tirosh and Tsamir, 1996).

What has not yet been investigated thoroughly is the extent to which a course in Cantorian Set Theory promotes prospective teachers' (i.e., prospective secondary school mathematics teachers') awareness of the need to avoid negating methods when comparing infinite sets, and what kind of course would be most efficacious for raising such awareness. Traditionally, Cantorian Set Theory is presented as a formal course, with little or no emphasis on students' intuitive tendencies to overgeneralize from finite to infinite sets. This paper describes another kind of course, where such tendencies are taken into account. A main aim of this paper is to assess the effects of such a course on prospective teachers' performance when comparing infinite sets.

The paper first describes the 'Enriching Course' in Set Theory, focusing on an activity that relates to the transition from finite to infinite sets (the course will also be referred to as the 'intuitive-based course'). Then the paper presents the research, which analyzed the influence of the intuitive-based course and the formal one, on prospective teachers' comparison of infinite sets. Prospective teachers' initial reactions (before taking any course of Cantorian Set Theory) to the comparisons of infinite sets tasks, were investigated in order to create a basis for judging the influences of both the 'formal' as well as the 'intuitively-based' courses. The paper concludes with a discussion and final conclusions.

## 2. THE ENRICHMENT COURSE IN SET THEORY

The Enrichment Course consisted of twenty-four weekly class sessions of 90 minutes each. The first five sessions were devoted to discussing connections between mathematics and reality, the axiomatic, independent nature of mathematical systems and the crucial role consistency plays in determining mathematical validity (referring to, for instance, Fischbein and Tirosh, 1996). Research findings regarding inconsistencies in students' mathematical performance, possible reasons for their occurrence and suitable teaching methods which are suggested in the literature, were discussed as well (e.g., Tall, 1990; Tirosh, 1990; Wilson, 1990).

The remaining nineteen sessions of the course related to Cantorian Set Theory, discussing defined and undefined concepts, axioms and theorems. Primarily, various finite and infinite sets, the null set, relations and operations between sets were presented. After comparing finite sets, infinite sets were compared and finally the powers of various sets were defined and discussed. The discussions of infinite sets followed Zermelo and Fraenkel's theoretical framework (using, for instance, Boolos, 1964/1983; Davis and Hersh, 1980/1990; Fraenkel, 1953/1961; Fraenkel and Bar-Hillel, 1958; Kitcher, 1947/1984; Smullyan, 1971); various teaching methods were applied, taking into account historical aspects and students' primary intuitions, and emphasizing the role of consistency in mathematics.

The activity described below was a part of the enrichment course. It describes the transition from finite to infinite sets. The presentation here consists of a description of the activity and comments regarding prospective teachers' reactions are included too.

### 2.1. *The activity – The transition from finite to infinite sets*

The aims of the activity were:

a. to develop prospective secondary mathematics teachers' abilities to critically reflect upon their ideas.

b. to increase prospective teachers' awareness of the characteristics that may be transferred and those which must be abandoned when passing from finite sets to infinite ones.
c. to convince prospective teachers that [1] consistency is the main tool of validation in mathematical theory; and [2] 1:1 correspondence is a suitable general method for the comparison of the number of elements in two infinite sets.

This activity was based upon findings of previous research related to students' reactions to comparisons of infinite sets tasks (e.g., Borasi, 1985; Tsamir, 1990; Tirosh, 1991). For instance, Tsamir and Tirosh (in press) found that students were strongly influenced by the way the problems were presented and tended to use different approaches to solve differently-presented, identical problems.

It was also found (e.g., Tirosh, 1991; Tsamir, 1990) that students used five methods to determine whether two infinite sets were equivalent: (1) all infinities are 'equal';[1] (2) incomparable – i.e., infinite sets are incomparable; (3) pairing – i.e., 1:1 correspondence; (4) inclusion; (5) intervals (i.e., when the elements of two sets have the same range but different intervals, then, the set in which the intervals are larger consists of fewer elements).

Accordingly, the problems created for the activity were designed so that different representations would trigger the use of different methods to compare the numbers of elements in infinite sets. This was meant to serve as a springboard for promoting prospective teachers' awareness of their methods of comparison, and for discussing the need to define one single method for comparing infinite sets.

### 2.2. *The Procedure*

The activity consisted of two main phases, the first deals with comparisons of finite sets and the second with comparisons of infinite sets. The first phase was subdivided into six stages, and the second phase into five stages. In every stage the prospective teachers either had to respond to a given assignment or reflect upon their previous responses. The concluding assignment in each of the two phases yielded a reflective summary of all the stages.

**Phase 1 – Comparing finite sets**

*Stage 1a – Assignment*

**Aim**

To trigger the generalization that comparisons of any pair of finite sets are done on the basis of one 'self evident' method: counting the number of elements in each set and comparing the two resulting numbers.

## Task

Here are two sets A and B:

A = {1, t, $\alpha$}  B = {7, w}

Is the number of elements in sets A and B equal? Yes / No

How did you reach this conclusion? ____________

Complete the following generalization:

In order to compare the number of elements in any two finite sets one could ____________

*Stage 1b – Assignment*

## Aim

To illustrate four methods for comparing the number of elements in two finite sets: pairing, inclusion, counting and intervals.

## Task

*Problem 1*

At a dance party all the students danced in couples, a boy and a girl in each couple. No pupils were left without a partner.

Z = {The boys}  W = {The girls}

Is the number of elements in set Z equal to the number of elements in set W ? Yes / No

How did you reach this conclusion? ____________

*Problem 2*

Given the sets:

X = { 1, 2, 3, 4, 5, 6, 7, 8, 9, 10, 11, 12}  Y = {a, b, c, d, e}

Is the number of elements in set X equal to the number of elements in set Y? Yes / No

How did you reach this conclusion? ____________

*Problem 3*

Given the sets:

Y = {a, b, c, d, e, f}  V = {a, b, c}

Is the number of elements in set Y equal to the number of elements in set V? Yes / No

How did you reach this conclusion? ____________

*Problem 4*

Dan was ill. The doctor prescribed one green tablet every 3 hours for the first week. Then, in the second week he was ordered to take a red capsule every 3 hours.

G = {The green tablets}  R = {The red capsules}

Is the number of elements in set G equal to the number of elements in set R? Yes / No

How did you reach this conclusion? ____________

*Problem 5*

Along the new promenade, a lane of trees was planted; a tree every 200 m. Every 400 m a street light was placed adjacent to a tree. The first and last trees had a street light next to them.

L = {The street lights}  T = {The trees}

Is the number of elements in set L equal to the number of elements in set T? Yes / No

How did you reach this conclusion? ____________

*Stage 1c – Reflection*

**Aims**

- To practice reflection upon the methods used to compare the number of elements in the sets.
- To conclude that comparisons are possible even when elements cannot be counted.
- To realize that counting, even when applicable, is not always the preferred method for comparing.
- To realize that counting is actually creating 1:1 correspondence between a set and a specific subset of the natural numbers.

**Task**

*Let's reflect –*

1. Which methods did you use in order to compare the given sets?

   ________________________________________

2. Which method did you use in each problem?

   ________________________________________

3. Could the number of elements be compared even when counting was impossible?

   ________________________________________

4. In cases that counting was applicable, was it always the preferable method?

   ________________________________________

5. What is the meaning of 'counting'?

   ________________________________________

**Comments**

All participants solved the problems in stage 1a, by counting. When asked to reach a generalized conclusion, they responded that *counting* is *the way* to solve problems concerning the comparison of finite sets.

The problems in Stage 1b were solvable by various methods. The participants did in fact use all four available methods to solve the problems. They listed these methods in response to the first assignment in the reflection-stage [1c] (counting – mainly in Problem 2; 1:1 correspondence – mainly in Problem 1; inclusion – mainly in Problem 3; and intervals – mainly in Problems 4 and 5). When all the methods used were presented in a single table, two findings emerged: (a) for each comparison, each prospective teacher used only a single method; (b) 1:1 correspondence was used by several prospective teachers to solve some problems.

The class discussion emphasized that: [1] the number of elements in two finitc sets could be compared even when counting was impossible (see Problems 1 and 5). [2] Even where counting was applicable, it was not necessarily the preferable method for attaining the solution (see Problems 3 and 4). [3] Counting is basically pairing, i.e., creating *1:1 correspondence* between the elements of a given set and a subset of the natural numbers.

*Stage 1d – Assignment*

**Aims**

- To increase prospective teachers' awareness of the validity of all four methods when comparing finite sets. Each of the methods where applicable, was valid – they could all be used alternatively, without risking conflicting answers.
- To increase prospective teachers' awareness of the extent to which each of the methods is applicable in a given problem in which finite sets are compared.

**Task**

Prospective teachers were asked to try to compare the sets in each of the five problems presented at Stage 1b, using all four methods (where possible) in each comparison.

*Try to apply all the four methods to each problem.*
*Sample Problem (Problem 1)*
At a dance, all the students danced in pairs, a boy and a girl in each couple.
No pupils were left without a partner.
Z = {The boys}    W = {The girls at assembly}
Is the number of elements in set Z equal to the number of elements in set W?

*In order to answer this question –*
Is '1:1' correspondence applicable? Yes / No
If your answer is Yes – Use this method to solve the problem –
Is the number of elements in set Z equal to number in set W? Yes / No

Is 'inclusion' applicable? Yes / No
If your answer is Yes – Use this method to solve the problem –
Is the number of elements in set Z equal to number in set W? Yes / No

Is 'intervals' applicable? Yes / No
If your answer is Yes – Use this method to solve the problem –
Is the number of elements in set Z equal to number in set W? Yes / No

Is 'counting' applicable? Yes / No
If your answer is Yes – Use this method to solve the problem –
Is the number of elements in set Z equal to number in set W? Yes / No

*Stage 1e – Reflection*

**Aims**

- To inculcate the role of consistency in mathematical systems.
- To demonstrate that there is one method which is always applicable – i.e., 1:1 correspondence.

**Task**

*Let's reflect –*

1. List the methods applicable to each problem.

| Problem No. | 1 | 2 | 3 | 4 | 5 |
|---|---|---|---|---|---|
| | Boys–girls | 1 to 16–a to e | a to f–a to e | tablets–capsules | lights–trees |
| a. Counting | | | | | |
| b. Pairing – 1:1 correspondence | | | | | |
| c. Inclusion | | | | | |
| d. Intervals | | | | | |

2. Is it OK to alternatively use these methods for the comparison of finite sets? Why?

________________________________________

3. Is there, after all, a preferable method?

________________________________________

4. How do the methods work?

________________________________________

5. Does the method of 'counting' enable to compare the numbers of elements in two given sets when there is no way to count?

   ________________________________________

   Does the method of 'inclusions' enable to compare the numbers of elements in two given sets when there is no inclusion relationship between the sets?

   ________________________________________

   Does the method of 'intervals' enable to compare the numbers of elements in two given sets when there is no way to predict the intervals?

   ________________________________________

   Does the method of '1:1 correspondence' enable to compare the numbers of elements in two given sets when there is no way to pair?

   ________________________________________

**Comments**

At Stage 1d the participants were asked to apply as many of the four available methods as possible for each problem, and at Stage 1e, to reflect on their actions. In the concluding class discussion they expressed awareness of the following:

When applicable, it is acceptable to use any of these methods for the comparison of finite sets, since their use never gives rise to negating answers. Consistency is always maintained. Consequently, the choice of a method for the comparison of finite sets is very much determined by personal inclination, which may be influenced by circumstantial factors such as the way the problem is represented, availability, and subjective convenience and preference. However, *1:1 correspondence* is always applicable. Determining the preferable method by its 'applicability' leads to the conclusion that *1:1 correspondence* is the preferable method for comparing the number of elements in finite sets. It has been found to be the only method that enables to compare the sets in both cases, when *a 1:1 correspondence* relationship between matching elements exists (indicating 'equality'), as well as in cases where it can be proved that there is no such relationship (indicating 'inequality').

*Stage 1f – Summary relating to finite sets*

**Aim**

To reexamine Stage 1a (Problem 1) while implementing the critical tools acquired after having performed the activities in stages 1b—1e, and drawing a final conclusion.

**Task**

*Final conclusion for finite sets*:

In order to compare the number of elements in any two finite sets you can

__________________________________________________

**Comments**

From the summaries yielded by Stage 1f it was clear that all participants accepted that: (a) the comparison of the number of elements in two finite sets can be conducted by using *counting*, *inclusion*, *intervals* and *1:1 correspondence* without violating the consistency of the theory, (b) the most applicable method is *1:1 correspondence*.

## Phase 2 – Comparing infinite sets

*Stage 2a – Assignment*

**Aim**

To illustrate that in comparing infinite sets, various methods are applied.

### Task

*Let's consider pairs of infinite sets*

*Problem I*

Given B = {3, 4, 5, 6, 7, 8, 9, ...}  W = {8, 9, 10, 11, ...}
The number of elements in sets W and B is equal / not equal.
Explain: ______________________________

*Problem II*

Given B = {3, 4, 5, 6, 7, 8, ...}  T = {3.1, 4.1, 5.1, 6.1, 7.1, ...}
The number of elements in sets T and B is equal / not equal.
Explain: ______________________________

*Problem III*

Given B = {3, 4, 5, 6, 7, 8, 9, ...}  P = {300, 400, 500, 600, 700, ...}
The number of elements in sets P and B is equal / not equal.
Explain: ______________________________

*Problem IV*

Given A = {1, 8, 27, 64, 125, 216, ...}  D = {3, 6, 9, 12, 15, 18, 21, ...}
The number of elements in sets A and D is equal / not equal.
Explain: ______________________________

*Problem V*

Given B = {5, 6, 7, 8, 9, ...}  M = {points on a given line}
The number of elements in sets M and B is equal / not equal.
Explain: ______________________________

*Stage 2b – Reflection*

**Aims**

- To promote awareness of the methods applied in comparing infinite sets.
- To consider which methods have been added and which have been lost in the move from finite to infinite sets.

### Task

*Let's reflect –*

1. What methods did you use to compare the number of elements in infinite sets?
2. What was added to the list of methods used for the comparison of finite sets and what had to be dropped from it? Why?

**Comments**

Participants used different methods for comparing the infinite sets given at Stage 2a. In the reflective task at Stage 2b they reported the use of 1:1 correspondence (Problem II); Inclusion (Problem I); All infinities are equal (Problem V); Intervals (Problem IV) and Incomparable (Problem V). The methods used for comparing both two finite and two infinite sets were *1:1 correspondence*, *inclusion* and *intervals*. It was emphasized that when moving from finite to infinite sets the methods added are: *all infinities are equal* and *incomparable*, while *counting* is no more applicable.

*Stage 2c – Assignment*

**Aim**

To promote awareness that the application of different methods when comparing *infinite* sets, leads to contradictory answers.

**Task**

*Try to compare the following sets using the provided methods*
*Sample Problem (Problem 3)*
Given
B = {1, 2, 3, 4, 5, 6, 7, . . .} P = {100, 200, 300, 400, 500, 600, 700, . . .}
The number of elements in sets P and B is equal / not equal.
Explain: ______________________________________

Is '1:1' correspondence applicable? Yes / No
If your answer is Yes – Use this method to solve the problem.
Is the number of elements in set B equal to the number of elements in set P? Yes / No

Is 'inclusion' applicable? Yes / No
If your answer is Yes – Use this method to solve the problem.
Is the number of elements in set B equal to the number of elements in set P? Yes / No

Is 'differences' applicable? Yes / No
If your answer is Yes – Use this method to solve the problem.
Is the number of elements in set B equal to the number of elements in set P? Yes / No

Is 'all infinities are equal' applicable? Yes / No
If your answer is Yes – Use this method to solve the problem.
Is the number of elements in set B equal to the number of elements in set P? Yes / No

*Stage 2d – Reflection*

**Aims**

- To promote prospective teachers' awareness of the contradictory answers yielded by different methods of comparing the number of elements in two *infinite* sets.
- To promote prospective teachers' awareness of the importance of consistency in mathematics.

- To conclude that only one method should be applied when comparing infinite sets.
- To show that 1:1 correspondence is the most applicable method when comparing infinite sets.

**Task**

*Let's reflect –*

1. Is it OK to alternatively use these methods, for the comparison of infinite sets? Why?
2. In your opinion, which (if any) of the various methods for comparing infinite sets is preferable?
   Why?

**Comments**

The responses to Stage 2c revealed, for instance, that sets B and P were taken to be simultaneously 'equal' and 'not-equal', as a result of using different methods in problem solving. The reflective assignment of Stage 2d led the participants to conclude that in the case of infinite sets an alternative use of different methods is impossible.

In the class discussion it was pointed out that each of the above mentioned methods is valid for the comparison of infinite sets, but only if used exclusively to compare all pairs of infinite sets. Otherwise, essential consistency is violated. Consequently, the choice of a method for the comparison of all infinite sets should be made in advance, and this method must then be used exclusively.

A question that naturally arose was: Is there a preferable method? i.e., one that provides more conclusive answers. Participants found that: (1) *incomparable* – allowed for no comparisons; (2) *all infinities are equal* – eliminated the reason for comparison for obvious reasons; (3) *inclusion* – was only occasionally applicable; (4) intervals – was only rarely applicable; and (5) 1:1 correspondence – was found to enable the comparison of infinite sets both when such a relationship existed (indicating 'equlity') and when it could be proved that there is no such correspondence (indicating 'inequality'). Moreover, this method was also applicable for comparing finite sets. Again 1:1 correspondence appeared to be the most applicable method.

*Stage 2e – Summing Up*

**Task**

Considering the various methods that were used to compare infinite sets, what happens in the extension from finite to infinite set-theory? ____________

**Comments**
The final assignment, before studying the theorems of the Cantorian Set Theory, was to sum up the characteristics gained and those lost in the transition from finite to infinite sets. Almost all participants pointed to the possibility of using various methods when comparing given sets as a characteristic lost in the transition from finite to infinite sets, the possible equivalency of a set and its proper subset was seen as a characteristic gained in this transition.

## 3. THE RESEARCH

The effects of the enrichment-course were assessed by comparing the performance of prospective teachers who participated in this course with that of prospective teachers who participated in a traditional, formal Cantorian Set Theory course. First, the research method will be described, and then the results will be presented and discussed.

### 3.1. *Method*

#### 3.1.1. *Subjects*
Three groups of prospective secondary school mathematics teachers studying in Israeli state teachers colleges, participated in the study. (a) Seventy one prospective teachers had never studied Cantorian Set Theory; (b) A hundred and ten prospective teachers had studied a 'formal course'of Cantorian Set Theory; and (c) A hundred and twenty five prospective teachers participated had studied in the 'enriching course' in Set Theory.

The formal course consisted of defined and undefined notions, axioms and theorems relating to infinite Set Theory, as presented by Zermelo and Fraenkel. In this course many comparison tasks were discussed, pointing at *1:1 correspondence* as being the method for conducting the comparisons within this theory. As mentioned before, the 'enriching course' put special emphasis on intuitive aspects.

The formal and the enrichment courses consisted of the same number of sessions, but were taught by two different instructors. While the instructor of the formal course was a specialist in mathematics (the subject matter), the person who taught the enrichment course specialized in mathematics education. The participants (those who participated in no course, those who studied the formal course and those who took the enrichment course) had a similar mathematical background, except for the Set-Theory course.

### 3.1.2. *Materials and procedure*

All participants were given a questionnaire that asked them to compare the numbers of elements in pairs of infinite sets and to justify their answers. The questionnaire is presented here.

### The questionnaire

*Compare the numbers of elements in the following pairs of sets and explain your answer*

| *The sets to compare* | | *Number of elements* |
|---|---|---|
| 1) A = {-4, -3, -2, -1, 0, 1, 2, 3, 4, . . .} | B = {1, 2, 3, 4, 5, 6, . . .} | equal / unequal |
| Explain your answer: ______ | | |
| 2) E = {-1, -2, -3, -4, -5, -6, . . .} | B = {1, 2, 3, 4, 5, 6, . . .} | equal / unequal |
| Explain your answer: ______ | | |
| 3) I = {2, 4, 6, 8, 10, . . .} | J = {1, 4, 9, 16, 25, . . .} | equal / unequal |
| Explain your answer: ______ | | |
| 4) D = {1/n \| n is a natural number} | K = {0.*3*xyt. . . \| all decimal fractions with 3 as a 'tenth' digit} | equal / unequal |
| Explain your answer: ______ | | |
| 5) B = {1, 2, 3, 4, 5, 6, . . .} | M = {points on segment ST<br>S ______ T} | equal / unequal |
| Explain your answer: ______ | | |
| 6) T = {points on a straight line} | M = {points on a 7 cm line segment} | equal / unequal |
| Explain your answer: ______ | | |
| 7) G = {points on a circle. R = 7 cm} | H = {points on a circle R = 10 cm} | equal / unequal |
| Explain your answer: ______ | | |

All participants were given about 90 minutes to answer the questionnaire, and about ten of each group were interviewed, in order to get a better insight into their ideas. All participants who took either a formal course or an enrichment course were given the questionnaire about two months after the courses ended. No task presented in the questionnaire had been discussed with the participants who took the enrichment course.

## 3.2. *Results: The effects of the enrichment course and the formal course*

This section relates to judgments, justifications and attention to consistency of the participants in the enrichment course and the formal course.

### 3.2.1. *The 'equal/unequal' judgments*

It seems that prospective teachers *who had not studied Cantorian Set Theory did not have a global grasp regarding the equality of infinite sets and approached each problem separately.* Certain problems intuitively triggered 'equal' judgments (e.g., about 62% of the judgments regarding the comparison of the set of natural numbers with the set of points on a given line segment), while other problems intuitively triggered ‘unequal’ judgments (e.g., about 88% of the judgments relating to the comparison of the set of natural numbers with the set of whole numbers greater than –5). This intuitive approach to problem solving was not necessarily consistent with the mathematical line of reasoning.

*The rate of ‘equal’ judgments, given by prospective teachers who had studied the formal course exceeded 47% on each problem.* Taking into account the fact that two problems described unequal sets, one may assume that participants occasionally chose their judgements by chance.

A survey of the judgments with reference to their validity within the framework of Cantorian Set Theory, showed the following (see Table I)[2]. Prospective teachers who had studied the formal course yielded a higher percentage of valid judgments than those who had participated in no course. However, the highest rate of valid judgments was presented, in all cases, by those who had participated in the enrichment course, which focused on intuitive aspects. For instance, more than 70% of those who studied the formal course, and almost all who had studied the enrichment course, gave valid judgments to the problems presenting equivalent sets (both $\aleph_0$ and c). The judgments given by participants who studied no course on problems with equivalent sets varied. While most of them identified the equivalency of the set of natural numbers with the set of negative whole numbers, the very same participants failed to grasp the equivalency between the set of natural numbers and the set of whole numbers greater than –5.

### 3.2.2. *The justifications*

Both ‘equal’ and ‘unequal’ judgments were justified in various ways, which could be classified into eight types. Five were designated previously in this paper: *1:1 correspondence, all infinities are equal, inclusion, intervals,* and *infinite sets are incomparable*. The two additional methods were: *bounded vs. infinite* – claiming that sets that are somehow bounded (e.g., the number of points in a given segment which is bounded, say, in terms of its length) must have less elements than sets that have no such bounds (e.g., all the natural numbers.); and *power* – justifying equal / unequal judgments by referring to the (in)equality of powers of the sets (‘the infinite numbers’ of their elements or their magnitudes); sometimes even specifying whether the relevant power was $\aleph_0$ or c.

The eighth type of justifications – marked as *Other ideas* in Table I related to non-explanatory, irrelevant ideas, such as *‘Finite ideas’* – i.e., relating to the sets as finite and not infinite sets regardless of the instructions given; *‘Content interpretations’* – i.e., relating to the numerical value of the elements presented and neglecting the quantity of elements in the sets;

TABLE I

Frequencies (in percent) of judgements and justifications to each problem

| | Equivalent sets – $\aleph_0$ | | | | | | | | | Equivalent sets – c | | | | | | Unequivalent sets | | | | | |
|---|---|---|---|---|---|---|---|---|---|---|---|---|---|---|---|---|---|---|---|---|---|
| | {-4,-3,-2,-1,0,1...} {1,2,3,4} | | | {-1,-2,-3,-4,...} {1,2,3,4,...} | | | {2,4,6,8,10,...} {1,4,9,16,25,...} | | | {points-7 cm circle} {10 cm. circle} | | | {points-segment} {points-line} | | | {1/n: n is Natural} {0.3xyt...} | | | {1,2,3,4,5,6,...} {points on S- - - - -T} | | |
| | N-ST | F-ST | E-ST | N-ST | F-ST | E-ST | N-ST | F-ST | E-ST | N-ST | F-ST | E-ST | N-ST | F-ST | E-ST | N-ST | F-ST | E-ST | N-ST | F-ST | E-ST |
| n = | 63 | 104 | 113 | 69 | 104 | 120 | 69 | 105 | 113 | 65 | 92 | 112 | 66 | 95 | 112 | 63 | 95 | 111 | 59 | 96 | 105 |
| EQUAL | 12 | 66 | 97 | 95 | 98 | 100 | 33 | 73 | 97 | 57 | 80 | 99 | 49 | 72 | 96 | 62 | 47 | 16 | 59 | 53 | 14 |
| 1-1 corres. | – | 17 | 45 | 78 | 56 | 54 | 8 | 24 | 49 | 2 | 7 | 57 | – | 9 | 46 | – | – | – | – | – | – |
| Power | – | 33 | 47 | – | 27 | 46 | – | 32 | 46 | – | 32 | 41 | – | 29 | 49 | – | – | – | – | – | – |
| W. 1-1 corres. | – | – | 3 | – | – | – | – | 1 | – | – | 2 | 1 | – | 2 | – | – | 12 | 5 | 10 | 17 | 9 |
| W. power | – | – | 2 | – | 1 | – | – | 1 | – | – | 4 | – | – | 1 | 1 | – | 7 | 9 | – | 14 | 5 |
| $\infty = \infty$ | 9 | 15 | – | 15 | 14 | – | 22 | 15 | 1 | 55 | 34 | – | 49 | 30 | – | 62 | 28 | 2 | 44 | 20 | – |
| Incomparable | 2 | – | – | – | – | – | – | – | – | – | – | – | – | – | – | – | – | – | 5 | 1 | – |
| Other | 1 | 1 | – | 2 | – | – | 3 | – | 1 | – | 1 | – | – | 1 | – | – | – | – | – | 1 | – |
| UNEQUAL | 88 | 34 | 3 | 5 | 1 | – | 67 | 27 | 3 | 43 | 20 | 1 | 51 | 28 | 4 | 38 | 53 | 84 | 41 | 47 | 86 |
| 1-1 corres. | – | – | – | – | – | – | – | – | – | – | – | – | – | – | – | – | 3 | 10 | – | 1 | 12 |
| Power | – | – | – | – | – | – | – | – | – | – | – | – | – | – | – | – | 37 | 73 | – | 25 | 72 |
| W. 1-1 corres. | – | 2 | 2 | – | – | – | – | 3 | 1 | – | – | – | – | 1 | 1 | – | – | 1 | – | – | – |
| W. power | – | 4 | – | – | – | – | – | 1 | 1 | – | 5 | – | – | 6 | 2 | – | – | – | – | – | 1 |
| Inclusion | 86 | 28 | 1 | – | – | – | 15 | 10 | – | 41 | 15 | 1 | 36 | 16 | 1 | 5 | 3 | – | 7 | 7 | – |
| Intervals | – | – | – | – | – | – | 47 | 13 | 1 | – | – | – | – | – | – | – | 3 | – | – | – | – |
| Infinite bounded | – | – | – | – | – | – | – | – | – | – | – | – | 9 | 5 | – | 22 | 7 | – | – | 1 | – |
| Incomparable | 2 | – | – | 2 | – | – | 2 | – | – | 2 | – | – | 2 | – | – | 2 | – | – | 2 | – | – |
| Other | – | – | – | 3 | 1 | – | 3 | – | – | – | – | – | 4 | – | – | 9 | – | – | 32 | 13 | 1 |

*'It depends on'* – i.e., attempting to compare the elements by ascribing arbitrary attributes to them (e.g., when comparing the number of points in two segments: 'It depends on the size of the points in each segment.); and *No response*.

3.2.2.a. *The tendency to use 1:1 correspondence or power*
Since the frame of reference for this research was Cantorian Set Theory, the justifications were first validated within this context. The only valid methods for establishing the equality or inequality of any two infinite sets are either *1:1 correspondence* or *power*. *1:1 correspondence* is the fundamental method, which was used both by prospective teachers who had and who had not studied Cantorian Set Theory. The method of *power*, which is rooted in Cantor's theory of infinite numbers and based on the method of *1:1 correspondence*, was not used, and could not be expected to be used, by prospective teachers who had not previously studied this theory (see Table I). The findings indicate that the highest rate of valid justifications (i.e., a correct use of either *1:1 correspondence* or *power)* was attained, in all cases, by those who had participated in the enrichment course. However, prospective teachers who had not studied Set Theory usually did not offer valid justifications for their comparisons. They presented no *power* justifications, and rarely did they provide *1:1 correspondence* justifications.

Comparison of the number of elements in the set of the natural numbers to the number of elements in the set of negative whole numbers yielded the highest rates of valid justifications for all participants (about 80% of those who had taken no course, 85% of the formal course students and all the enrichment course students). While the comparison of non-equivalent sets triggered the lowest rates of valid justifications for formal course and enrichment course participants (about 25–40% and 80–85%, respectively).

It is noteworthy that a number of prospective teachers used the correct ideas of *1:1 correspondence* and *power* incorrectly (see W-1:1 corres., and W-Power in Table I). For instance, a participant who had taken the formal course said: 'The points on a straight line and those on a certain line segment have the power c. A straight line is also known to be a number-line, so each number relates to a single point. Thus there is *1:1 correspondence* between the points of the line and the set of natural numbers'. A participant in the enrichment course, incorrectly claimed that the number of rational numbers 1/n was equal to the number of decimals 0.*3*xyt. . ., explaining that: '*According to the rules of Set Theory both sets have the same power*'.

Such incorrect uses of *1:1 correspondence* or *power* considerations, indicated prospective teachers' *intention* to compare by the methods acceptable in the Cantorian context. In recording prospective teachers' justifications (Table I) both correct as well as incorrect use of *1:1 correspondence* and *power* were listed, since one's declared tendency to perform by the valid methods was found important in itself. Moreover, when dealing with aspects of consistency the use of only *1:1 correspondence* and *power* (no matter whether in a valid or an invalid manner), were appreciated as a possible inclination to preserve consistency. Thus, when analyzing prospective teachers' consistent behavior valid and invalid use of these methods were viewed similarly.

3.2.2.b. *The use of global justifications*

Two kinds of 'global' methods were presented in the comparison of infinite sets: *'All infinite sets are equal' (infinity = infinity)* and *'Infinite sets are incomparable' (incomparability).* One would expect that this type of approach would force those who use it to consistently answer all problems. Hence, when examining the table that lists the frequencies of justifications to each problem (Table I) we would expect to find the same rate of *infinity = infinity* or *incomparable* justifications for all problems. However, this is not the case: on the whole, the *infinity = infinity* justification was used much more frequently than the *incomparable* justification. The *infinity = infinity* justification was used by about 10–60% of no-course participants, about 14–35% of formal course participants and almost never by enrichment course participants. The same holds true for the few no-course participants who presented *incomparability* as justification. Moreover, these participants even provided equal or unequal judgments and justified these judgments by stating *'Infinite sets are actually incomparable.'* This leads us to conclude that not all these prospective teachers felt necessarily bound by these methods.

3.2.3. *Consistent and inconsistent use of justifications*

This section extends the examination, briefly mentioned before, of the consistency of the approach by which prospective teachers conducted the comparison tasks. For this purpose a number of logically interrelated methods were combined, leaving four categories of methods: (a) extended *1:1 correspondence* – a unification of valid and invalid *1:1 correspondence* and power; (b) *extended inclusion* – a unification of *inclusion, intervals* and *bounded-infinite*; (c) *infinity = infinity* and (d) *incomparable*.

The survey of prospective teachers' consistent behavior examined their tendency to compare all pairs of infinite sets by one single, unified method (Tables II and III)[3] and analyzed their tendency to consistently use global methods (Table IV).[4]

TABLE II

Frequencies (in %) of prospective teachers using various numbers of unified methods when comparing infinite sets

| | N-ST (n = 69) | F-ST (n = 108) | E-ST (n = 125) |
|---|---|---|---|
| No of unified methods used | | | |
| 1 | 7.0 | 37.5 | 93.9 |
| 2 | 27.5 | 43.1 | 5.3 |
| 3 | 64.0 | 18.5 | 0.9 |
| 4 | 1.4 | 0.9 | – |

TABLE III

Frequencies (in %) of prospective teachers using methods as a single method for comparing infinite sets

| | N-ST (n = 69) | F-ST (n = 108) | E-ST (n = 125) |
|---|---|---|---|
| The only unified method used | | | |
| 1:1 Correspondence | – | 37.5 | 93.9 |
| Only *valid* use of 1:1 corres or power part of the unified 1:1 correspondence | – | 10.9 | 31.6 |
| All infinities are equal | 4.2 | 2.8 | – |
| Incomparable | 2.8 | 0.9 | – |

Table II shows that the frequency of the use of a single method increases from no-course students (7%), through formal-course students (about 38%), to enrichment-course students (about 94%). That is, almost all prospective teachers who had followed no Set Theory course, a substantial number of the formal-course participants and only a few enrichment-course students used more than one method when comparing infinite sets. Further, using three unified methods to compare all pairs of infinite sets was the most frequent number of methods used by no-course prospective teachers, two methods were most frequently used by the formal-course participants, while almost all the enrichment-course participants used a single (unified) method for their comparisons.

TABLE IV

Frequencies (in%) of negating answers

| | N-ST (n = 71) | F-ST (n = 110) | E-ST (n = 125) |
|---|---|---|---|
| The negating arguments: | | | |
| 'Incomparable' & 'Equal / Unequal' | 5.7 | 2.7 | – |
| 'Not Equal' & 'All infinite sets are equal' | 73.2 | 40.9 | 1.6 |

In the next stage we examined nature and frequency of choices of only one method. Table III shows that the no-course participants, when following only a single method, used only global methods – i.e., *infinity = infinity* or *incomparability*. However, these methods are not only invalid within the Cantorian frameworks, the prospective teachers also used them quite inconsistently. Formal-course participants, when using a single method, usually chose *1:1 correspondence* or *power*, sometimes in a valid but mostly in an invalid manner. A very small number used either *infinity = infinity* or *incomparable* as their methods. The highest rate of use of one method – always that of the *unified 1:1 correspondence* (in a valid or invalid manner) was found among the enrichment course participants.

One of the givens when comparing infinite sets is that the use of more than one (unified) method will eventually lead to contradiction. Participants who had not taken a Set Theory course, probably could not be aware of this and had difficulties becoming aware of this, since they treated each problem separately and did not reach any global conclusions regarding infinite sets.

However, two types of contradictions seem to be especially extreme – i.e., [1] declaring that infinite sets are incomparable and then proceeding to compare them; and [2] stating that all infinite sets are equal (have the same number of elements) and then proceeding to provide 'unequal' as a solution to infinite comparison tasks. While the first type was quite rare the second was found in almost 75% of the no-course participants, 40% of the formal-course participants, but in less than 2% of the enrichment-course participants (see Table IV).

To summarize, prospective teachers who had not studied any course in Cantorian Set Theory (1) tended to overgeneralize from comparison of finite sets to comparison of infinite ones; (2) were not aware of inconsistencies that occurred as a result of the overgeneralization, and (3) neglected the demand to always use a single method. These tendencies decreased among participants who had studied the formal course, and were almost non-existent among enrichment-course participants. The latter exhibited a mindful, critical ability in reducing overgeneralization.

## 4. Discussion and Final Conclusions

In line with other research regarding the responses of students of various age levels when comparing the powers of infinite sets (e.g., Fischbein, Tirosh and Hess, 1979; Martin and Wheeler, 1987; Tirosh and Tsamir, 1996), our findings indicate that prospective secondary school mathematics teachers who had followed no Set Theory course intuitively compared the numbers of elements in *infinite* sets in a manner which was only valid for the comparisons of *finite* ones. That is, in most cases these students used different methods while examining each problem *separately*, thus neglecting the inevitable consequences of the resulting incompatibilities.

This phenomenon of viewing each problem as a new, independent issue, while ignoring existing connections and referring to no linking frameworks, was described by Vinner (1990) as the 'compartmentalization' of knowledge. Vinner claimed that compartmentalization can be a cause of inconsistencies, and explained that:

> By 'compartmentalization', I refer to situations in which two pieces of knowledge (or information) that are known to an individual and that should be connected in the person's thought process nevertheless remain unrelated.
>
> (Vinner, 1990, pp. 92)

Compartmentalization of mathematical knowledge might also be the reason for the tendency of prospective teachers who took no Set Theory course to come up with contradictory claims, e.g., (1) stating 'all infinite sets are equal' as well as 'the number of elements in any given two sets is unequal'; or (2) stating in one case that 'infinite sets are incomparable', while, in another case, concluding that the number of elements in two sets is either 'equal' or 'unequal'.

By and large, usually no-course participants neither followed a single method for the comparison of infinite sets, nor did they view this as necessary. Both their practical comparisons as well as their suggested general approach for comparing infinite sets expressed their major concern for the ability to solve such complicated mathematical tasks. Pre-occupied with finding 'a way' to compare each pair of infinite sets, they tended to accept the method that came to their minds and then legitimized it by reference to the results they attained.

The two methods most frequently applied by prospective teachers who had followed no course, were *infinity = infinity* and *inclusion,* while *1:1 correspondence* was rarely considered. Still, the high rate of use of the latter method was rather outstanding when comparing the number of natural numbers with that of the whole, negative numbers. The visual aspects of the specific task probably influenced these prospective teachers. When presented with the sets {1, 2, 3, 4, 5, . . .} vs. {-1, -2, -3, -4, -5, . . .} they intuitively tended to match the visually, similar pairs of elements: 1 with (-1), 2 with (-2), 3 with (-3) n with (-n) etc. (see, for instance, in Duval, 1983; Tirosh and Tsamir, 1996). However, no notion of *1:1 correspondence* arose, for instance, in the comparison of the sets {1, 2, 3, 4, 5, . . .} and {-4, -3, -2, -1, 0, 1, 2, 3, 4, 5 . . .}. Almost all no-course participants used *inclusion* considerations, while the others reasoned that *all infinities are equal* or that *infinities are incomparable*.

It should be taken into account that prospective teachers who had followed no course, had no formal instruction related to Cantorian Set Theory to guide them towards using *1:1 correspondence* as the sole method for their comparisons. However, one may expect prospective teachers to infer that in the transition from finite to infinite sets, much like in other extensions of number systems which they have previously experienced, some characteristics will be gained while others lost. Lacking the formal knowledge as to the nature of these characteristics, consistency would be the only tool available for prospective teachers' reflective criticism of their own mathematical claims. In this respect, our findings validated that prospective teachers' intuitive grasp of infinite sets was an overgeneralization of their grasp of finite ones, usually with no attention given to the consistency of their responses.

On the other hand, prospective teachers who had taken either formal or enrichment courses, had been introduced to the formal, theoretical Cantorian framework and to the use of *1:1 correspondence*. Therefore, it was only natural that, these participants used *1:1 correspondence* when comparing infinite sets and often chose it as the single method for all comparisons, much more frequently than did no-course participants. Contrary to the latter, those who had studied Set Theory tended to identify *1:1 correspondence* (and *power*) as the only preferable method for comparing infinite sets, and occasionally justified this claim in terms of consistency. Still, there was conclusive evidence of the continued influence of intuitive ideas, which repeatedly interfered (mostly) with the formal-course participants' ability to criticize their comparisons and be aware of contradictions.

The present findings clearly indicate that, the highest rate of success in comparing the number of elements in infinite sets was found among prospective teachers who had taken the enrichment course. In addition, these students were most consistent in their use of a single method, and in the 'reflection stages' of the activity they expressed awareness of the need to preserve consistency within a given mathematical system.

The question that arises is, why did the enrichment course have a much more impressive impact on prospective teachers' mathematical performance than the formal course? After all, both courses included an extensive informative presentation of the relevant formal theorems, proofs and definitions; both had offered substantial opportunities to exercise the major method of determining the power (equivalency) of infinite sets. In the enrichment course, however, intuitive aspects of the mathematical performance were discussed, sometimes at the expense of formal training. Taking into account learners' primary intuitive ideas in the course of their mathematical studies led both to high percentages of correct comparisons of infinite sets and to awareness of the need to assure consistency (e.g., Fischbein, 1983, 1987; Papert, 1980; Tall, 1980).

When students begin the study of a new subject, they usually apply some intuitive knowledge. These primary intuitions are generally rooted in everyday life and previous practical experience. Included in this 'intuitive' knowledge may also be knowledge acquired in previously studied mathematical systems- and this may lead to the attempt to apply notions valid in one mathematical system to another in which they are not valid. Research in the field of mathematics education has indicated that primary intuitions often interfere with students', and prospective teachers', performance in many mathematical fields (e.g., Ball, 1990; Fischbein, 1987, 1993; Tall, 1990; Tall and Vinner, 1981; Tirosh, 1991).

The accumulated knowledge about students' ways of thinking about infinity was taken into consideration when building the outline for the enrichment course of Set Theory. In class, the instructor was constantly aware of and attuned to these intuitions and the way they affect prospective teachers' ways of thinking and performing. Participants were repeatedly reminded of the role of intuitive ideas in their mathematical performance and that this occasionally results in inconsistencies. They were therefore guided to relate to consistency as a crucial tool for determining validity within any given mathematical system. Of course, awareness of intuitions alone is not enough. Formal knowledge of Cantorian Set Theory is essential as well. In this awareness of the need for consistency and its link to formal knowledge may lie the key to why in the enrichment course, although it devoted less time to teaching formal knowledge, the final results were better.

Finally, when teaching mathematics courses, instructors should be attentive to the relations among formal and intuitive knowledge and to the conflicts which may arise in the mismatching applications of these different types of knowledge. This is true of courses dealing with Set Theory, but it holds equally for any course dealing with mathematical systems. Thus the cycle of building and rebuilding a course is an 'infinite' one.

## NOTES

1. The phrase 'infinite sets are equal' is used to indicate that *the infinite sets are equivalent* or that *the infinite sets have the same number of elements.*
2. Table I relates only to the participants who presented judgments, ignoring in each problem those who omitted the specific item. Therefore the number of subjects listed in Table I is lower than the number reported previously in the text. The labels: N-ST, F-ST, E-ST refer to the 'participants who had not taken a Set Theory course', those who had taken a Formal Course and those who studied the Enrichment Course, respectively.
3. Tables II and III relate only to the participants who presented justifications, ignoring those who omitted justifications to *all* problems. Therefore the numbers of subjects listed in these Tables differ from the numbers noted in Table I and are lower than the numbers reported previously in the text.
4. Table IV relates to those among *all* subjects, who at least once presented a pair of 'negating arguments'.

## REFERENCES


Almog, N.: 1988, *Conceptual Adjustment in Progressing from Real to Complex Numbers: An Educational Approach,* Unpublished thesis for the Master's degree. Tel Aviv University, Tel Aviv, Israel, (in Hebrew).

Ball, D. L.: 1990, 'Prospective elementary and secondary teachers' understanding of division', *Journal for Research in Mathematics Education* 21(2), 132–144.

Boolos, G.: 1964/1983, 'The iterative concept of set', in P. Benacerraf and H. Putman (eds.), *Philosophy of Mathematics*, Cambridge University Press, Cambridge, pp. 486–502.

Borasi, R.: 1985, 'Errors in the enumeration of infinite sets', *Focus on Learning Problems in Mathematics* 7, 77–88.

Davis, P. J. and Hersh, R.: 1980/1990, *The Mathematical Experience*, Penguin, London, pp. 136–140, 161–162.

Duval, R.: 1983, 'L'obstacle du dedoublement des objects mathematiques', *Educational Studies in Mathematics* 14, 385–414.

Falk, R., Gassner, D., Ben Zoor, F. and Ben Simon, K.: 1986, 'How do children cope with the infinity of numbers?' *Proceedings of the 10th Conference of the International Group for the Psychology of Mathematics Education*, London, England, pp. 7–12.

Fischbein, E.: 1983, 'The role of implicit models in solving elementary arithmetical problems', *Proceedings of the 7th Conference of the International Group for the Psychology of Mathematics Education*, Rehovot, Israel, pp. 2–18.

Fischbein, E.: 1987, *Intuition in Science and Mathematics*, D. Reidel, Dordrecht, The Netherlands.

Fischbein, E.: 1993, 'The interaction between the formal and the algorithmic and the intuitive components in a mathematical activity', in R. Biehler, R. W. Scholz, R. Straser and B. Winkelmann (eds.), *Didactic of Mathematics as a Scientific Discipline*, Kluwer, Dordrecht, The Netherlands, pp. 231–345.

Fischbein, E., Jehiam, R. and Cohen, D.: 1995, 'The concept of irrational numbers in high-school students and prospective teachers', *Educational Studies in Mathematics* 29(1), 29–44.

Fischbein, E. and Tirosh, D.: 1996, *Mathematics and Reality*, unpublished manuscript, Tel Aviv University, Tel Aviv, Israel (in Hebrew).

Fischbein, E., Tirosh, D. and Hess, P.: 1979, 'The intuition of infinity', *Educational Studies in Mathematics* 10, 3–40.

Fischbein, E., Tirosh, D. and Melamed, U.: 1981, 'Is it possible to measure the intuitive acceptance of a mathematical statement?' *Educational Studies in Mathematics* 12, 491–512.

Fraenkel, A. A.: 1953/1961, *Abstract Set Theory*, North-Holland, Amsterdam.

Fraenkel, A. A. and Bar-Hillel, Y.: 1958, *Foundations of Set Theory*, North-Holland, Amsterdam.

Greer, B.: 1994, 'Rational numbers', in T. Husen and N. Postlethwaite (eds.), *International Encyclopedia of Education* (Second ed.), Pergamon, London.

Hart, K.: 1981, *Children's Understanding of Mathematics, 11–16,* Murray, London.

Hefendehl, H. L.: 1991, 'Negative numbers: Obstacles in their evolution from intuitive to intellectual constructs', *For the Learning of Mathematics* 11(1), 26–32.

Kitcher, P.: 1947/1984, *The Nature of Mathematical Knowledge*, Oxford University Press, pp. 101–148.

Klein, R. and Tirosh, D.: 1997, 'Teachers' pedagogical content knowledge of multiplication and division of rational numbers', *Proceedings of the 21st Conference of the International Group for the Psychology of Mathematics Education,* Lahti, Finland, 3, 144–151.

Martin, W. G. and Wheeler, M. M.: 1987, 'Infinity concepts among preservice elementary school teachers', *Proceedings of the 11th Conference of the International Group for the Psychology of Mathematics Education,* France, pp. 362–368.

Moloney, K. and Stacey, K.: 1996, 'Understanding decimals', *Australian Mathematics Teacher* 52(1), 4–8.

Papert, S.: 1980, *Mindstorms: Children, Computers and Powerful Ideas*, Harvester, England.

Putt, I. J.: 1995, 'Preservice teachers ordering of decimal numbers: When more is smaller and less is larger!' *Focus on Learning Problems in Mathematics* 17(3), 1–15.

Smullyan, R. M.: 1971, 'The continuum hypothesis', in *The mathematical Sciences*, The M.I.T. Press, Cambridge, pp. 252–260.

Streefland, L.: 1996, 'Negative numbers: reflection of a learning researcher', *Journal of Mathematical Behavior* 15(1), 57–77.

Tall, D.: 1980, 'The notion of infinite measuring numbers and its relevance in the intuition of infinity', *Educational Studies in Mathematics* 11, 271–284.

Tall, D.: 1990, 'Inconsistencies in the learning of calculus and analysis', *Focus on Learning Problems in Mathematics* 12(3&4), 49–64.

Tall, D. and Vinner, S.: 1981, 'Concept image and concept definition in mathematics with particular reference to limit and continuity', *Educational Studies in Mathematics* 12, 151–169.

Tirosh, D.: 1990, 'Inconsistencies in students' mathematical constructs', *Focus on Learning Problems in Mathematics* 12, 111–129.

Tirosh, D.: 1991, 'The role of students' intuitions of infinity in teaching the cantorial theory', in D. Tall (ed.), *Advanced Mathematical Thinking*, Kluwer, Dordrecht, The Netherlands, pp. 199–214.

Tirosh, D. and Tsamir, P.: 1996, 'The role of representations in students' intuitive thinking about infinity', *International Journal of Mathematics Education in Science and Technology* 27(1), 33–40.

Tsamir, P.: 1990, *Students' Inconsistent Ideas about Actual Infinity*, Unpublished thesis for the Master's degree. Tel Aviv University, Tel Aviv, Israel (in Hebrew).

Tsamir, P. and Tirosh, D.: 1992, 'Students' awareness of inconsistent ideas about actual infinity', *Proceedings of the 16th Conference of the International Group for the Psychology of Mathematics Education*, Durham, USA, 3, 90–97.

Tsamir, P. and Tirosh, D.: 1994, 'Comparing infinite sets: intuitions and representations', *Proceedings of the 18th Conference of the International Group for the Psychology of Mathematics Education* Lisbon, Portugal, 4, 345–352.

Tsamir, P. and Tirosh, D.: 'Consistency representations: The case of actual infinity', *Journal for Research in Mathematics Education*, in press.

Vinner, S.: 1990, 'Inconsistencies: Their causes and function in learning mathematics', *Focus on Learning Problems in Mathematics* 12(3&4), 85–98.

Wilson, P.: 1990, 'Inconsistent ideas related to definition and examples', *Focus on Learning Problems in Mathematics* 12(3&4), 31–48.

*School of Education,*
*Tel Aviv University,*
*69978 Tel Aviv,*
*Israel*
*and*
*Mathematics Education,*
*Kibbutzim Teachers College,*
*62507 Tel Aviv,*
*Israel*

RUHAMA EVEN

# INTEGRATING ACADEMIC AND PRACTICAL KNOWLEDGE IN A TEACHER LEADERS' DEVELOPMENT PROGRAM

ABSTRACT. This study examines an attempt to encourage integration of knowledge learned in the academy with knowledge learned in practice as a means to challenge educational practitioners' – teacher leaders and inservice teacher educators – existing conceptions and beliefs, and promote intellectual restructuring. The article centers on two components of the Manor Program for the development of teacher leaders and educators. The first component focuses on expanding academic knowledge, by helping the participants become acquainted with studies on students' and teachers' conceptions and ways of thinking in mathematics. The second component focuses on the integration of knowledge learned in the academy with knowledge learned in practice by conducting a mini-study.

## 1. INTRODUCTION

There is an accumulation of evidence that many teachers do not promote meaningful learning of mathematics. This is due to various factors, among them, the conceptions and beliefs that teachers and teacher leaders and educators hold about teaching and learning – for example, a conception of learning as transfer of knowledge from the teacher to the student (Strauss and Shilony, 1994) – which perpetuate problematic ways of teaching. Due to the dialectic of beliefs and practice (e.g., Cobb, Wood and Yackel, 1990), in that change in one domain is connected to change in the other, aiming at intellectual restructuring by challenging teachers' and teacher leaders' conceptions and beliefs about teaching and learning has the potential to contribute to making a meaningful change in teaching.

Research indicates that experience in the classroom by itself is not enough to challenge and encourage re-examination of conceptions and beliefs. Desforges (1995) claims that humans in practice tend to close down rather than open up to experience by means of intellectual restructuring. In his synthesis of research in this area, Desforges concludes that experiencing anomalous events in the classroom does not usually encourage teachers to restructure their own conceptions and beliefs about teaching and learning. Rather, according to Chinn and Brewer (1993), when encountering contradictory data in practice, teachers usually ignore them,

*Educational Studies in Mathematics* **38:** 235–252, 1999.

reject, exclude as irrelevant, hold in abeyance, re-interpret them in terms of their existing personal theory, or make minor and peripheral changes to the theory. Consequently, practice alone is not enough to challenge educational practitioners' existing conceptions and beliefs.

Genuine integration of knowledge learned in the academy with knowledge learned in practice has the potential to challenge educational practitioners' existing conceptions and beliefs, and support intellectual restructuring, because such integration, as described by Leinhardt, McCarthy Young and Merriman (1995), 'involves examination of the knowledge associated with one location while using the ways of thinking associated with the other location by asking learners to particularize abstract theories and to abstract principles from particulars' (p. 403). Such integration helps to make what is usually assumed and taken for granted, questionable and examinable.

An obstacle to such integration is the alienation that exists between practitioners in mathematics education and research. While research has flourished in the last decades, many practitioners (e.g., teachers, teacher leaders) are not familiar with it. This seems to be due, in part, to lack of practitioners' access to research (which is exasperated in countries, like Israel, since most of the research is published in English–a 'foreign' language), and also to a common belief among practitioners that academic studies are irrelevant to educational practice (e.g., Bromme and Tillema, 1995). This feeling of irrelevancy seems to stem from an unfulfilled expectation that 'research should provide reliable and relevant rules for action, rules that can be put to immediate use' (Kennedy, 1997, p. 10).

This article examines an attempt to make research in mathematics education meaningful for practitioners – teacher leaders and inservice teacher educators. It centers on making research more accessible to them, and on introducing ways in which research can be made relevant to them, as practitioners, even if it does not provide them with rules for action. This study is part of a more comprehensive study situated in the context of the *Manor Program*. The Manor Program aims to develop a professional group of teacher leaders and inservice teacher educators whose role is to promote teacher learning about mathematics teaching. Some of the program activities emphasize and build on the practical expertise of the participants, others introduce an academic perspective. In the following section I give a brief overview of the Manor Program, elaborating its academic component which is the focus of this article (for a comprehensive description of the multi-facets of the Manor Program, see Even, in press).

## 2. The Manor program

### 2.1. *Program focus*

The program emphasizes the following:

- The development of understanding about current views of mathematics teaching and learning.
- The development of leadership and mentoring knowledge and skills, and of work methods with teachers.
- The creation of a professional reference group.

The program centers on cognitive, curricular, technological, and social aspects of teaching different mathematical topics; it examines critical educational issues; it enhances mathematical knowledge; it emphasizes the development of leadership skills and methods for working with teachers; it encourages discussion of practical difficulties and dilemmas; and it focuses on initiating change in school mathematics teaching and learning.

### 2.2. *Operation of the program*

In this article, I focus on the first group of teacher leaders and inservice teacher educators who started the program in the 1993–94 academic year. The program extended over three years in an effort to allow sufficient time for the participants to learn, experience, and experiment with the topics and ideas encountered. Further, there was a need for development and growth in the participants' conceptions, beliefs, and dispositions about the nature of mathematics learning and teaching and about teaching teachers (Even, 1994). Such changes require time to become established.

During each school year, the participants met weekly for four hours with project staff and guest lecturers, either in whole-group sessions or in parallel teams. In addition, they conducted weekly two-hour professional development activities, some explicitly focused on initiating change in mathematics teaching and learning. As an overall assignment for each year, the participants prepared portfolios that documented their learning experiences. They received feedback on partial drafts several times throughout the year, both from project staff and from their peers.

### 2.3. *Academic component*

Part of the program was devoted to developing the value participants attached to inquiry into student learning of different topics in mathematics, and into student and teacher conceptions and ways of thinking. We wanted the participants to look at mathematics learning 'from the student point of view', to examine what might be the meaning of the widespread constructivist claim (supported by numerous cognitive studies) that students' ideas are not necessarily identical to the structure of the discipline nor to what was intended by instruction; that students construct and develop their own knowledge and ideas about the mathematics they learn.

Current research and professional rhetoric (e.g., Barnett, 1991; Even and Markovits, 1993; Fennema et al., 1996; National Council of Teachers of Mathematics, 1991; Rhine, 1998) recommends that attention be paid to students' mathematics learning and thinking in teacher professional development programs as a means to support teacher learning to promote students' meaningful learning and understanding of mathematics. In a program such as the Manor Program, such focus can serve as a vehicle to challenge the participants' own conceptions about learning and teaching mathematics, as well as a means to introduce ideas to be explored in the teacher development activities they themselves conduct.

Because the course participants were meant to lead, teach and work with mathematics teachers, we also wanted them to be sensitive to teachers' conceptions and ways of thinking in mathematics and mathematics teaching, and to look at mathematics teaching 'from the teacher point of view'. As teachers were once students, it is reasonable to assume that teachers' knowledge and students' knowledge are not dichotomized; that teachers, like students, develop and construct their own ideas about the mathematics they have learned (and the mathematics they are still learning), which, in part, is the mathematics that they teach – ideas which may be different from what is sometimes assumed or expected. Indeed, several studies support this claim (e.g., Ball, 1990; Even, 1993; Tirosh and Graeber, 1990). However, it is not as well accepted nor appreciated as the similar claim about students.

This component of the program seemed a natural place for focusing on research, deepening the academic background of the participants in mathematics education (the participants received graduate credit for this advanced academic component), and encouraging integration of knowledge learned in the academy with knowledge learned in practice. In contrast with the approach of the Cognitively Guided Instruction (CGI) – a founder project that focuses on changing teachers' beliefs and practices by helping teachers acquire research-based knowledge about students' thinking – we did not provide the Manor Program participants with explicit research-based models of children's thinking in specific mathematical topics. Research on student thinking at the level of junior- and senior-high school mathematics does not seem to support the existence of such models. Rather, similar to the Integrating Mathematics Assessment (Rhine,

1998) and the Mathematics Classroom Situations (Even and Markovits, 1993; Markovits and Even, in press) approaches, we focused on presenting the program participants with research-based key features of student and teacher thinking in different mathematical topics. Thus we aimed at challenging and expanding the participants' understanding of students' and teachers' ways of making sense of the subject matter and the instruction.

A large part of this component of the program included reading, presentations and discussions of research articles on students' and teachers' conceptions and ways of thinking in mathematics. Later, the participants were asked to choose one of the studies presented in the course, replicate it (or a variation of it) with students and with teachers, and then write a report describing the subjects' ways of thinking and difficulties, comparing their results with the original study. This article examines the first group of the Manor Program participants' encounters with research, as a means for intellectual restructuring, and concentrates on an examination of the nature of these experiences.

## 3. Methodology

### 3.1. *Participants*

The first group of program participants consisted of 30 mathematics educators who were selected from approximately 100 applicants. Selection was based on the following criteria: (a) a first degree either in mathematics or in a mathematics-related field, (b) experience in mathematics teaching and inservice work with mathematics teachers, at least one of which in grade nine or above, (c) agreement to conduct weekly inservice work with a group of secondary mathematics teachers during the program, (d) reputation as a successful teacher with the potential to become a good teacher leader or teacher educator, and (e) a reasonable spread of participants across the country.

Participants' teaching experience varied from 5 to 29 years with a mean of 18 years. About one fourth of the participants taught at the junior-high level only, another fourth at the senior-high level only, and the rest had teaching experience at both levels. About two-thirds of the participants held only a bachelors' degree, one-third had a masters' degree, and one participant a Ph.D.

About one-half of the participants had less than five years experience in conducting inservice work with secondary teachers or as mathematics coordinators in their schools, almost one-half had between five to ten years of experience, and a few (three) had more than 15 years of experience.

Almost all had participated in many inservice courses on mathematics teaching; about half had also completed courses for school coordinators or courses dealing with leadership skills although the leadership courses were not specific to mathematics or mathematics education.

### 3.2. *Data collection and analysis*

Several types of data were collected in the larger study. Here I mention the ones used for this study:

*Questionnaires.* For a quick impression of the participants' feelings about the academic component of the program, we used three short Likert-type response scale questionnaires. In the first (Q1), which was administered two months after the program began, the participants rated their interest in, and the importance of, the last seven whole-group sessions, three of which were devoted to research in mathematics education. In the second questionnaire (Q2), which was administered at the end of the first year, the participants rated the contribution of six topics dealt with in the first year whole-group sessions, of which three were on research. Finally, in the third questionnaire (Q3), which was administered at the end of the second year, the participants rated their interest in, and the contribution of, eight topics dealt with in the second year whole-group sessions, of which four centered on research.

*Interviews.* The questionnaires provided important yet limited information. In order to understand better how the participants perceived the research sessions in the program, a sample of the participants was also interviewed; 13 at the end of the first year (I1), ten at the end of the second (I2), and nine at the end of the third year (I3). In these semi-structured interviews they were asked to refer to the various program components and reflect on their learning experiences. The interviews were audio-taped and later transcribed. For the purpose of this study references to the academic component only were analyzed.

*Observations.* All the whole-group sessions and several of the sessions conducted in parallel teams, were observed, some were video-taped. These observations provided additional information on the general atmosphere, participants' interests, dispositions and difficulties, and the development of ideas.

*Informal talks.* Numerous informal talks with the participants occurred, either in the participants' initiation or the researcher's. In these talks the participants asked questions, consulted, expressed their feelings about various program components, and reflected on their experiences in the program. The issue of research in mathematics education was central in these talks.

*Project staff meetings*. The discussions and consultations during the biweekly staff meetings provided additional information and a forum for discussing developing ideas and analysis relevant to this study.

*Portfolios*. As an overall assignment for each year, the participants prepared portfolios that documented their learning experiences. For the purpose of this study only the second year reports on the mini-studies they conducted were analyzed.

## 4. ACQUAINTANCE WITH RESEARCH IN MATHEMATICS EDUCATION

To help the participants become familiar with relevant research literature, a large part of the whole-group sessions included presentations and discussions of research articles that focus on student and teacher mathematical conceptions and thought processes. The participants also read related articles. Most of the studies presented in the course dealt with students' conceptions, and only a few with teachers' conceptions, which reflects the present in the math education literature. The main issues raised and discussed were closely connected to secondary school mathematics:

- discrepancies between concept image and concept definition of function (e.g., Even, 1993; Vinner and Dreyfus, 1989),
- difficulties in translating and making connections between different representations of function (e.g., Bell and Janvier, 1981; Even, 1998),
- the objective and subjective multi-facets of variable (e.g., Arcavi and Schoenfeld, 1988; Küchemann, 1981),
- cognitive development in algebra (e.g., Sfard, 1995),
- students' conceptions of the derivative (e.g., Amit and Vinner, 1990),
- 'proofs that explain and proofs that prove' (e.g., Hanna, 1990),
- hypotheses and proofs in technological environments (e.g., Hershkowitz and Schwarz, 1995),
- various levels and aspects of geometrical thinking (e.g., Hershkowitz, 1990),
- conceptions of irrational numbers and $\pi$ in particular (e.g., Tall and Schwarzenberger, 1978),
- statistical thinking in a technological environment (Ben-Zvi and Friedlander, 1997).

In addition we dealt with

- classroom cultures that support and promote the development of mathematical reasoning (e.g., Collins, Brown and Newman, 1989; Lampert, 1990).

Analysis of the Likert-type response scale questionnaires provided important information on how the academic component of the program was perceived by the participants. It indicates that, from the beginning of the program, the participants felt that the sessions devoted to studies in mathematics education were interesting and important. These results do not explain, of course, what was interesting or important, but they suggest a 'proof of a concept' which was crucial for the development of the Manor Program; that learning about research in mathematics education can interest practitioners.

The interviews and observations of course sessions provided additional support for this conclusion and shed some light on what was interesting or important. Many participants explicitly stated, on their own initiative, during and after related whole-group sessions, how astonished they were to learn that students 'are able to think that way'. When asked during an interview (I2) to relate to the whole-group sessions, some added that learning about research helped them make explicit things they already 'felt' but did not have the language to express. These reactions suggest two kinds of learning. One has to do with developing appreciation of the idea that students construct their knowledge in ways which are not necessarily identical to the instruction. The other has to do with conceptualizing and making explicit naive and implicit knowledge.

Whereas interest in learning about students' conceptions and ways of thinking seemed high from the beginning, the participants did not seem to have the same initial interest in learning about teachers' knowledge. This became apparent from the discussions during the course meetings, which concentrated almost exclusively on students' conceptions, even when several studies that dealt with teachers' subject-matter knowledge were presented. Also, when the 13 first-year interviewees were asked in I1 what they had learned in the course, all mentioned the issue of students' conceptions and ways of thinking, but only one mentioned that she had also learned about *teacher* thinking. Thus we concluded that this issue deserved special attention, as part of developing their role as teacher leaders and educators.

More than that, in our aim at intellectual restructuring related to both students and teachers, we felt that reading and discussing research were not enough. Taking into account the current dichotomy between theory and practice, it would be unreasonable to believe that such acquaintance with research would be enough to make a difference. In line with the model proposed by Leinhardt et al. (1995), we assumed that the synthesis of theoretical and practical sources of knowledge is more complete and informative than either one alone. We wanted the participants to examine theoretical knowledge acquired from reading and discussing research in the light of their practical knowledge. Vice-versa, we wanted the participants to build upon and interpret their experience-based knowledge using research-based knowledge. The following section describes the use of a mini-study as a means to address, both the issue of lack of focus on teacher knowledge, and the issue of integrating knowledge learned in the academy with knowledge learned in practice.

## 5. CONDUCTING MINI-STUDIES

### 5.1. *Reluctance to conduct a mini-study*

On several occasions during the first year, in an attempt to encourage the participants to examine the more theoretical knowledge in the light of their practice, the project staff suggested the participants choose a few of the research tasks presented to them in various articles and course presentations, to try them out with their students and/or teachers, and to examine if they obtained similar results. Very few followed this suggestion, and the reasons given usually centered on time limitations, not having access to a suitable student population, and not feeling comfortable with the idea of investigating their colleagues' understanding of the mathematics they teach. These reasons are not all of the same nature. The first two are technical, and may indicate real, objective difficulties. But they may also reflect lack of interest, and that such an activity is irrelevant to them. The third response reflects a problematic perception of their role as teacher leaders and educators, viewing exploration of teacher knowledge not as an integral part of what they need to deal with. It also highlights the problematic work conditions of teacher leaders and educators – teachers do not usually feel uncomfortable about investigating their students' knowledge and understandings. More on the issue of investigating teacher conceptions is presented later.

Because we believed that conducting a mini-study was worthwhile we suggested it more formally, as one of three options for the main task at the end of the first year portfolio. Only three out of the 30 participants chose this option. When asked, most claimed that the other two topics seemed more interesting and that it seemed useless to replicate a study that had already been done by someone else. Such a response suggests a view of research as 'truth' – a problematic view which may inhibit the development of productive connections between research and practice.

The fact that so few chose to replicate a study, while so many exhibited a genuine interest in learning about research in the formal course sessions, seems to reflect a complex alienation between them as practitioners and academic research. Consequently, it was decided to make the 'mini-study' compulsory and it was assigned as one of the main tasks to be included in the second-year portfolios. The participants had to choose one of the studies presented in the course, and replicate it or a variation of it with students and teachers. Intellectual restructuring depends on deep processing of experiences (Desforges, 1995), which is more likely to occur if the activity requires personal involvement and having to present the ideas and reasoning to others (Chinn and Brewer, 1993). Therefore, the participants were required to write a report that would describe the subjects' ways of thinking and difficulties, and compare the results with the results of the original study.

### 5.2. *Integrating knowledge learnt in the academy with knowledge learnt in practice*

Almost all of the participants felt that they learned a great deal from doing the task. They referred to two kinds of benefit. One kind is academic. They felt that replicating a study expanded their theoretical knowledge, and that it helped them to develop better understanding of the issues presented and discussed in the articles they read. For example, Sarah (a pseudonym, as are all the names in this article) said:

> When you read a research article, it is one level of depth. When you have to re-do it and implement it again, it is another level. I mean, what I know now about the study, about its hypothesis, its findings, and the theoretical material, I certainly wouldn't know after reading it once or twice or even if I had summarized it – it is much more. It became mine.

It seems that re-doing a mini-study with real students and teachers provided opportunities for examining theoretical matters by particularizing them in a specific context.

The other kind of benefit clearly reflects an integration of knowledge learned in the academy with knowledge learned in practice. It involves learning about *real* students and teachers in a situation relevant to their practice. Almost all of the participants chose to start from a study that was in some way relevant to their actual work in the field. They used the original study as a base for developing a study which would help them answer questions that were important to them about their students or the teachers with whom they work. Alma and her colleague, for example, based their mini-study on Lee and Wheeler's (1986) study, which examines the kinds of justification that students apply to their judgments of the truth or falsity of propositions in an algebraic situation. As a rationale for choosing this study they explained (in the introduction to their mini-study report in the second year portfolio) that the topic of generalizations and justifications in algebra was presented during the course from different angles, which made them curious about how their own students would behave in such problem situations. They especially wanted to learn to what extent algebraic tools are used by students for these purposes, because when they inspected current textbooks they found that they do not do enough to develop inductive reasoning and the need to justify generalizations. In her interview (I2) Alma explained that the results of their study (which was conducted with 7th and 8th grade students and teachers) were so interesting that she decided, in addition to the course requirements, to 'compare students from 2nd grade up to grades 7 and 8'. This comparison helped her 'understand what is going on in elementary school'.

Sarah summarizes nicely this kind of benefit in her interview (I2):

> In a mini-research, in contrast to an article which is completely theoretical, you have question marks about the findings. Could it be like this? Is it only a coincidence that this happened? Will it happen to students I know? My students? It is very interesting to see what really happens. To duplicate the study and see, to support the original findings or refute them ...

Thus conducting a mini-study with real students seemed to encourage the participants to examine their experience-based knowledge in light of their theoretical knowledge. The results of this examination were surprising for many of the participants – but for two opposing reasons. Sarah's interview (I2) excerpt illustrates one kind of surprise. She explains how she found out that the students could do much more than she expected:

> Even though I have worked for 30 years as a teacher, I was surprised by some of the things that we found in the group of students we studied. The students reached much higher levels of thinking than what I would have given them credit for. So it was very interesting.

Dalia's interview (I2) excerpt illustrates another kind of surprise. As opposed to Sarah, she points out how she wanted to 'prove' that her students would do better than the ones in the original study, because she (and the other teachers in her school) finished teaching them about irrational numbers just several months before the study was carried out. However, she was amazed:

> We did a replication of the lecture [a study on student conceptions of irrational numbers presented in one of the sessions]. Simply, I was amazed by the results. I said, well, this is a topic that we deal with in grade 9. It was several months after we had taught the material. And I said, OK, no problem. Our students, for sure, would know better than those students at the university [in the original study]. And we were shocked that actually with us it was the same as there. That was the interesting part. We also shared with the senior-high teachers on this topic. This was an interesting part.

As we can learn from such responses, integrating knowledge learned in the academy with knowledge learned in practice enabled even experienced caring teachers, like those who participated in the program, to learn important things about teaching mathematics and about their students. They learned that what they thought they knew about their students was not necessarily a good representation of their knowledge and abilities. Depending on their background and the specific project they chose to work on, some learned that what seemed to them to be too difficult can actually be dealt with successfully by their students. Others found that even well-planned teaching may not produce the kind of learning they expected, because learning processes do not necessarily mirror instruction.

As the last excerpt above illustrates, some of the participants further connected theory and practice, and discussed their findings with other teachers in their school. This issue is elaborated in the following section.

### 5.3. *Further connecting theory to practice: Working with other teachers*

Before conducting the mini-studies, several of the interviewees remarked, during formal sessions or informal talks, that they wanted to help their colleagues at school become aware of the theoretical material to which they had been introduced in the course, but they were not sure how to go about it. After they had learned first-hand about real students' conceptions in a specific topic, had developed their knowledge and understanding in various domains, and had become more skillful in work with other teachers, several participants decided to work in the same direction in their weekly teacher development meetings. At first, the teachers were not very interested in learning about research, and they objected to the apparently non-practical nature of the activity – an attitude which gradually changed, as the following quotation from Sharon's interview (I3) illustrates. 'At first the teachers objected: 'It's a waste of time. Instead, one can prepare another worksheet.' But today they look forward to the exposure to research and articles.'

Indeed, a number of the participants started to be aware of and appreciate the possibilities theoretical knowledge offered as a context for teachers to study their practice. This appreciation is illustrated in Sharon's description of how she directed the attention of the teachers in her school to topics that seemed important to her, how they analyzed the curriculum differently as a result of newly acquired theoretical knowledge, and how her school's practice was changing.

> Actually, our objective is to change some elements in the actual teaching. To introduce elements that may be familiar to some of the teachers but the solutions of how to implement and bring them to work in the field are not well-known. For example, one of the studies that I worked on was on the concept of the algebraic expression as a variable. I was exposed to it last year and I brought it [to the staff]. And it was simply amazing, the students' responses. And then we gave these things to the class and we raised additional questions which are actually already in the textbook. But [this time] we concentrated on them and therefore the

students gained some more . . . So we need to apply what we learn in the field. Because if there is no impact in the field then there is no point in studying this research.

Sharon emphasizes in her report that, for her, learning about research is worthwhile only if it can be connected to her practice. The above excerpt exemplifies one way she constructs this connection. She chooses an issue that is relevant in the context of mathematics teaching in her school – in the above case, the meaning of algebraic expressions. This is a central issue in the junior-high mathematics curriculum in Israel. Becoming acquainted with research in this area (e.g., Jensen and Wagner, 1981; Sfard, 1991), Sharon expanded her understanding of the dual nature of algebraic notations: processes and objects. For example, that the expression $2x + 6$ stands both for the process 'add two times x and six' and for an object. She also learned that students often conceive algebraic expressions as a process only. Based on this theoretical knowledge, Sharon designed and conducted several meetings with the teachers in her school. She helped the teachers to become aware of the phenomenon, they then analyzed the textbooks they were using with respect to their potential to develop a structural approach to algebraic expressions, and finally, planned their teaching in this direction. Noticing a change in student learning was an important factor in this endeavor.

### 5.4. *Awareness and understanding of teachers' ways of thinking*

In general, even though they were required to study students and teachers, the participants paid much more attention to the 'student part' in their study than to the 'teacher part'. In many cases, the project staff needed to remind them that the mini-study should also include teachers. Even when teachers were included in the study from the beginning, in many cases they were not a natural integral part of the work. For example, in a research proposal on 'The algebra before the algebra', Gil and Daniel chose to replicate a study on generalizations and justifications based on Arcavi, Friedlander and Hershkowitz (1990). In the theoretical background, the two gave a fairly comprehensive overview of research on children's related conceptions of generalizations and justifications before and during the study of algebra. Then, they presented their research question: Does the study of algebra improve students' ability to generalize and justify? Notice that teachers were not mentioned at all at this point. However, the population included grade 7 students (who have not yet studied algebra), grade 8 and 9 students (who have already studied some algebra), and, without any warning, inservice and pre-service teachers. No intrinsic explanation was given for the inclusion of teachers in the study. After specifying the research population (of which teachers were a part), the research instruments were presented. They, again, included a questionnaire and individual interviews with students only. Part of the written feedback from the project staff on this research proposal included the questions: 'Why do you investigate teachers? What do you want to learn from this?' Gil and Daniel were also advised to add relevant material to their theoretical background.

In a draft of their paper, Gil and Daniel explained that they investigated students who have not formally studied algebra and students who are at different stages of their formal algebra studies, to obtain a comparison between the populations and thus answer their research question. However, they added (irrelevant to their previous description) that they also investigated what junior-high school teachers think about students' ways of justification and proof in grades 7, 8, and 9. Further, the research instruments included a description of the questionnaires and the interviews used with students while teachers were not mentioned. The results section did include a description of the teachers' anticipation of students' difficulties in responding to the questionnaire.

Part of the written feedback from the project staff referred to this weakness in the paper: 'Why is it important [to study what teachers think]? How are teachers related [to this work]?' And even more explicit: 'Gil and Daniel! What do researchers say about the importance of 'what teachers think' on 'what students do'?' After the research instruments were specified, Gil and Daniel were asked: 'How about the teachers? What were the respective research instruments?' And, at the discussion section, where teachers' anticipation of their students' responses to the questionnaire tasks were discussed, the project staff remarked: 'This is the place where you can say more about the relation between what teachers thought would happen and what really happened, and also try to explain it.'

In the final version of the paper the inclusion of teachers seemed like a natural component of the study. In their explanation of the purpose of their study, Gil and Daniel specified, as before, the student and teacher population. But then added that they decided to include teachers in the study because:

> We think that teacher awareness of their students' ways of generalization and justification, at the beginning stage of formal learning of algebra, would influence their work. It would influence their lesson planning and their teaching. They would create classroom situations where every child would have opportunities to go through processes of justification and generalization.

The research instruments referred to teachers as well. Gil and Daniel explained that the teachers responded to the same questionnaire tasks as the students did, but the teachers were asked about their anticipation of their students' difficulties. The discussion section in the final version of the paper also included a comparison between teachers' anticipation of their students' responses and what Gil and Daniel found about the students in their study.

Informal talks with Gil and Daniel indicated that the changes described above in the various versions of their written work were neither artificial nor superficial. These were actual changes in their understanding of the purpose of learning about teachers' conceptions.

## DISCUSSION

Part of the Manor Program centered on deepening and expanding the participants' understanding about students' and teachers' conceptions and ways of thinking in mathematics. It was included in the program to support their learning to foster teacher learning to promote student learning. In this article we have examined two main components of this part of the Manor Program. The first focuses on expanding academic knowledge, by helping the participants become acquainted with research in mathematics education through presentations, readings and discussions of research articles on students' and teachers' mathematical conceptions and ways of thinking. The second component focuses on the integration of knowledge learned in the academy with that learned in practice by conducting a mini-study.

The findings indicate that learning in the academy can be designed to be interesting and relevant to practitioners – the teacher leaders and educators who participated in the program. The first component of the program expanded participants' understanding of the notion that students construct their own knowledge and ideas about the mathematics they learn, in ways which are not necessarily intended by instruction. Also, acquaintance with research in mathematics education via discussion of research articles, supported the development of what were initially intuitive, naive and implicit ideas, into more formal, deliberated, solid and explicit knowledge.

The second component of the program – performing a mini-study – supported the first component and expanded it. For example, the first component contributed to learning in general that students construct their own knowledge. The mini-study made this general theoretical idea more specific, concrete and relevant for the participants. They learned what the constructivist view might mean in a practical context. For example, some examined specific cases that exemplified that learning processes are complicated, no matter how 'clear' the instruction. Others examined cases that exemplified that, against expectation, students are able to deal with sophisticated mathematical ideas. In addition, the mini-studies provided opportunities to focus on issues which otherwise would have been neglected, like teacher knowledge.

Writing the report served as a means for deep processing which is associated with intellectual restructuring. In this it has similar characteristics to action research, in which teachers choose and then explore some aspect of their own teaching, and later make public their exploration. Desforges (1995) claims that because of the teachers' personal involvement, and their knowledge that judgment and reasoning will have to be communicated – factors that lead to deep processing – action research is more reliably associated with intellectual restructuring and revised practices than are other forms of teachers' professional development.

Furthermore, writing the report assisted in genuine integration of knowledge learned in the academy and that learned in practice. For the participants to articulate what they had learned from the mini-study, and to present it in a written form, was, in a way, what Leinhardt et al. (1995) describe as a transformation of knowledge learned in one location into forms associated with another. The participants needed to transform knowledge learned in practice into forms usually associated with the academy, and vice versa. This process of transformation is neither easy nor trivial. But it provides opportunities for reflective and analytic thought that leads to intellectual restructuring via knowledge integration.

This study investigates intellectual restructuring and change in knowledge and beliefs, but not actual change in teaching practice. The latter is beyond the scope of this study. Still, cases such as Sharon's school, where acquaintance with theoretical knowledge contributed to an actual change in teaching practice and student learning, suggest that emphasis on integrating academic knowledge and practical knowledge in the ways described in this article are a promising approach, not only for developing teachers leaders and educators, but also in teacher education. The importance of this integration seems to be rooted in its potential to offer new ways of examining and understanding practice. It can assist, as Leinhardt et al. (1995) describe, to make situational, intuitive, and tacit practical knowledge more formal and explicit, and theoretical knowledge more available for use in practice.

## REFERENCES

Amit, M. and Vinner, S.: 1990, 'Some misconceptions in calculus: Anecdotes or the tip of an iceberg?', in G. Booker, P. Cobb and T.N. de Mendicuti (eds.), *Proceedings of the 14th International Conference for the Psychology of Mathematics Education*, Program Committee, Mexico, Vol. 1, pp. 3–10.

Arcavi, A., Friedlander, A. and Hershkowitz, R.: 1990, 'The algebra before algebra', *Mesarim-Misparim (Journal for Mathematics Teachers in Israel)*, 3(3), 48–62 (in Hebrew).

Arcavi, A. and Schoenfeld, A.H.: 1988, 'On the meaning of variable', *Mathematics Teacher* 81, 420–427.

Ball, D.L.: 1990, 'Examining the subject matter knowledge of prospective mathematics teachers', *Journal for Research in Mathematics Education* 21(2), 132–143.

Barnett, C.: 1991, 'Building a case-based curriculum to enhance the pedagogical content knowledge of mathematics teachers', *Journal of Teacher Education* 42(4), 263–272.

Bell, A. and Janvier C.: 1981, 'The interpretation of graphs representing situations', *For the Learning of Mathematics* 2(1), 34–42.

Ben-Zvi, D. and Friedlander, A.: 1997, 'Statistical thinking in a technological environment', in J.B. Garfield and G. Burrill (eds.), *Proceedings of the International Association for Statistical Education Round Table Conference*, International Statistical Institute, Voorburg, The Netherlands, pp. 45–55.

Bromme, R. and Tillema, H.: 1995, 'Fusing experience and theory: The structure of professional knowledge', *Learning and Instruction* 5(4), 261–267.

Chinn, C.A. and Brewer, W.F.: 1993, 'The role of anomalous data in knowledge acquisition: a theoretical framework and implications for science instruction', *Review of Educational Research* 63(1), 1–49.

Cobb, P., Wood, T. and Yackel, E.: 1990, 'Classrooms as learning environments for teachers and researchers', in R.B. Davis, C.A. Maher and N. Noddings (eds.), *Constructivist Views on the Teaching and Learning of Mathematics*, National Council of Teachers of Mathematics, Reston, VA, pp. 125–146.

Collins, A., Brown, J.S. and Newman, S.E.: 1989, 'Cognitive apprenticeship: Teaching the crafts of reading, writing, and mathematics', in L.B. Resnick (ed.), *Knowing, Learning, and Instruction*, Lawrence Erlbaum, Hillsdale, NJ, pp. 453–493.

Desforges, C.: 1995, 'How does experience affect theoretical knowledge for teaching?', *Learning and Instruction* 5(4), 385–400.

Even, R.: 1993, 'Subject-matter knowledge and pedagogical content knowledge: prospective secondary teachers and the function concept', *Journal for Research in Mathematics Education* 24(2), 94–116.

Even, R.: 1994, '*Tel-Aviv Project for Improving Mathematics Teaching in Junior-High Schools: A Teacher Development Approach (1990–1993)*', Unpublished manuscript, Weizmann Institute of Science, Rehovot, Israel (in Hebrew).

Even, R.: 1998, 'Factors involved in linking representations of functions', *Journal of Mathematical Behavior* 17(1), 105–121.

Even, R.: 'The development of teacher-leaders and in-service teacher educators', *Journal for Mathematics Teacher Education*, in press.

Even, R. and Markovits, Z.: 1993, 'Teachers' pedagogical content knowledge of functions: Characterization and applications', *Journal of Structural Learning* 12(1), 35–51.

Fennema, E., Carpenter, T., Franke, M., Levi, L., Jacobs, V. and Empson, S.: 1996, 'A longitudinal study of learning to use children's thinking in mathematics instruction', *Journal for Research in Mathematics Education* 27(4), 403–434.

Hanna, G.: 1990, 'Some pedagogical aspects of proof', *Interchange* 21(1), 6–13.

Hershkowitz, R.: 1990, 'Psychological aspects of learning geometry', in P. Nesher and J. Kilpatrick (eds.), *Mathematics and cognition*, Cambridge University Press, Cambridge, UK, pp. 70–95.

Hershkowitz, R. and Schwarz, B.B.: 1995, *Reflective Processes in a Technology-Based Mathematics Classroom*, Paper presented at the annual meeting of the American Educational Research Association, San-Francisco.

Küchemann, D.E.: 1981, 'Algebra', in K.M. Hart (ed.), *Children's Understanding of Mathematics*, John Murray, London, pp. 102–119.

Jensen, J.R. and Wagner, S.: 1981, 'Three perspectives on the process uniformity of beginning algebra students', in S. Wagner (ed.), *Proceedings of the Second Annual Meeting of the PME-NA*, Athens, GA, pp. 133–139.

Kennedy, M.M.: 1997, 'The connection between research and practice', *Educational Researcher* 26(7), 4–12.

Lampert, M.: 1990, 'When the problem is not the question and the solution is not the answer: Mathematical knowing and teaching', *American Educational Research Journal* 27(1), 29–63.

Lee, L. and Wheeler, D.: 1986, 'High school students' conception of justification in algebra', in G. Lappan and R. Even (eds.), *Proceedings of the Seventh Annual Meeting of the PME-NA*, East Lansing, MI, pp. 94–101.

Leinhardt, G., McCarthy Young, K.M. and Merriman, J.: 1995, 'Integrating professional knowledge: The theory of practice and the practice of theory', *Learning and Instruction* 5(4), 401–408.

Markovits, Z. and Even, R.: in press, 'Mathematics classroom situations: in-service course for elementary school teachers', in B. Jaworski, T. Wood and A.J. Dawson (eds.), *Mathematics Teacher Education: Critical International Perspectives*, Falmer Press, London.

National Council of Teachers of Mathematics: 1991, *Professional Standards for Teaching Mathematics*, Author, Reston, VA.

Rhine, S.: 1998, 'The role of research and teachers' knowledge base in professional development', *Educational Researcher* 27(5), 27–31.

Sfard, A.: 1991, 'On the dual nature of mathematical conceptions: Reflections on processes and objects as different sides of the same coin', *Educational Studies in Mathematics* 22(1), 1–36.

Sfard, A.: 1995, 'The development of algebra: Confronting historical and psychological perspectives', *Journal of Mathematical Behavior* 14, 15–39.

Strauss, S. and Shilony, T.: 1994, 'Teachers' models of children's minds and learning', in I. Hirschfeld and S.A. Gelman (eds.), *Mapping the Mind: Domain Specificity in Cognition and Culture*, Cambridge University Press, Cambridge, pp. 455–473.

Tall, D.O. and Schwarzenberger, R.L.E.: 1978, 'Conflicts in the learning of real numbers and limits', *Mathematics Teaching* 82, 44–49.

Tirosh, D. and Graeber, A.: 1990, 'Inconsistencies in preservice teachers' beliefs about multiplication and division', *Focus on Learning Problems in Mathematics* 12, 65–74.

Vinner, S. and Dreyfus, T.: 1989, 'Images and definitions for the concept of function', *Journal for Research in Mathematics Education* 20, 356–366.

*Dept. of Science Teaching,*
*Weizmann Institute of Science,*
*Rehovot 76100,*
*Israel*
*E-mail: nteven@wiccmail.weizmann.ac.il*